Heavens Above!
A Binocular Guide
to the Southern Skies

Dedication:
To Marion
The true star in my life

Acknowledgements

In the preparation of this book from its inception, I have been encouraged and supported by my family and members of the Macarthur Astronomical Society.

Special thanks to John Rombi and Colin Howie for spending their valuable time in proof reading and improving the content of the book. In the end, any errors contained in this book are mine alone.

Heavens Above!

A Binocular Guide to the Southern Skies

Robert Bee

First published 2004 by Evalt
Second Edition 2006

Text, Cover Design, Illustrations by Robert Bee

© Robert Bee 2004, 2006

The right of Robert Bee to be identified as the author of this work has been asserted by him in accordance with the Copyright, Designs and Patents Act 1988.

This book is sold subject to the condition that it shall not, by way of trade or otherwise, be lent, resold, hired out, or otherwise circulated without the publisher's prior consent in any form of binding or cover other than that in which it is published and without a similar condition including this condition being imposed on the subsequent purchaser.
No part of this book may be reproduced for any purposes by any means except for permitted uses as specified under the Australian Copyright Act (as amended).

Bee, Robert 1946 -

Heavens Above!
A Binocular Guide to the Southern Skies

ISBN 1 876409 66 5

National Library of Australia Cataloguing-in-Publication entry
1. Southern sky (Astronomy) - Observers' manuals
2. Southern sky (Astronomy) - Amateurs' manuals
3. Binoculars
1. Title

522.0284

Printed in Australia by Snap Printing

CONTENTS

1. Introduction 7
2. What's to be seen with Binoculars? 9
3. What Binoculars are Best? 13
   - Aperture 13
   - Magnification 14
   - Exit Pupil 15
   - Field of View 16
   - Eye Relief 16
4. I See it, but What is it? 18
   - The Scale of Space 18
   - The Moon 20
   - The Planets 22
     - Mercury 23
     - Venus 24
     - Mars 25
     - Jupiter 27
     - Saturn 30
     - Uranus and Neptune 32
   - The Stars 32
   - Multiple Stars 33
   - Variable Stars 34
   - The Milky Way 34
   - Open Clusters 35
   - Globular Clusters 36
   - Nebulae 37
   - Galaxies 37
   - Constellations – Road Maps in the Sky 38
   - Star Names 39
   - Constellation Names and Abbreviations 40
   - Numbering Deep Space Objects 42
   - Messier Objects 42
   - Magnitudes of Stars 43
5. Practical Tips for Observing 45
   - Great Expectations 45
   - Dark Adaptation 46
   - Estimating Distance Angles 46
   - Tips for Steadying your Binoculars 47
   - Finding the South Celestial Pole 51
   - Averted Vision 54
6. A Tour of the Constellations 55
   - Visibility of Constellations 55
   - Using the Monthly Star Maps 57
   - **Monthly Star Maps** **58**

5

The Constellations:

Andromeda 82	Leo – The Lion 119
Antlia – The Air Pump 83	Lepus – The Hare 120
Apus – The Bird of Paradise 84	Libra – The Scales 121
Aquarius – The Water Carrier 85	Lupus – The Wolf 122
Aquila – The Eagle 86	Lyra – The Lyre 123
Ara – The Altar 87	Monoceros – The Unicorn 124
Auriga – The Charioteer 88	Musca – The Fly 126
Bootes – The Herdsman 89	Norma – The Set Square 127
Cancer – The Crab 91	Ophiuchus – The Serpent Holder 128
Canes Venatici–The Hunting Dogs . 93	Orion – The Hunter 130
Canis Major – The Greater Dog 95	Pavo – The Peacock 132
Capricornus – The Sea Goat 97	Pegasus – The Winged Horse 133
Carina – The Keel 98	Perseus 134
Centaurus – The Centaur 100	Pisces – The Fishes 136
Cetus – The Whale 103	Puppis – The Stern 137
Chamaeleon – The Chameleon 104	Sagitta – The Arrow 139
Coma Berenices - Berenice's Hair 105	Sagittarius – The Archer 140
Corona Australis – The S'th'n Cr'n 106	Scorpius – The Scorpion 142
Corona Borealis – The N'th'n Cr'n 107	Sculptor – The Sculptor 144
Crux – The Southern Cross 108	Scutum – The Shield 145
Cygnus – The Swan 109	Serpens – The Serpent 146
Delphinus – The Dolphin 111	Taurus – The Bull 147
Dorado – The Goldfish 112	Telescopium – The Telescope ... 149
Equuleus – The Little Horse 113	Triangulum – The Triangle 150
Gemini – The Twins 114	Triangulum Australe-S'th'n Trgl . 151
Grus – The Crane 115	Tucana – The Toucan 152
Hercules 116	Vela – The Sails 153
Hydra – The Water Snake 118	Vulpecula – The Fox 155

SECTION 1.

Introduction – The Joy of Star Gazing

There is no doubt that the Southern Hemisphere has the best view of our Galaxy. That's not a parochial boast but a scientific fact.

Only from the Southern Hemisphere can you easily see into the heart of our Galaxy, the Milky Way.

Only from the Southern Hemisphere can you see the two best Globular Clusters, Omega Centauri and 47 Tucanae.

Only from the Southern Hemisphere can you see the most glorious of Open Clusters, the Jewel Box.

We are indeed fortunate here Down Under to be blessed with such a great night sky. Even more so that in such a sparsely populated continent such as Australia, large light polluting cities are few and far between, giving rise to more easily found sites for that most precious of prizes – a Dark Sky.

If you already live in a small country town or on the land, you'll know exactly what I mean. And if, as a city dweller, you've ever driven out of your city at night onto an unlit rural road and have bothered to take your eyes off the road's centre line and momentarily looked upwards, you'll also know.

The stars are out there in their thousands, just waiting to be admired.

Despite my years of amateur astronomy, I never fail to be struck dumb by the beauty of a clear, dark, country sky. One winter's evening, a work colleague and I were driving from Sydney to Orange. At about 7.30pm I was at the wheel and we were midway between Bathurst and Orange. We hadn't yet had dinner. On a long stretch of road, I noticed a particularly bright star dead ahead and played the 'What is it?" game. (It turned out to be Sirius.) My companion, possibly fearing for his life as I combined star searching and steering, suggested we pull over and have a good look. We did so, pulling into a truck rest stop area, turning off the car lights and stepping out into the cold.

It just so happened that I'd brought my binoculars with me, so we stood in the cold night, staring in awe at the unimaginable dark sky, splattered by the Milky Way and the surrounding constellations. It was only the cold and hunger that eventually tore us away from that sky.

As I drove the remaining distance to Orange, I resolved to write this book. That one inspiring hour by the roadside confirmed my opinion that there is a whole world of viewing to be had just with binoculars.

My companion who was never one for star gazing as he lived in a well lit northern Sydney suburb, was flabbergasted at what he had seen and vowed to buy a pair of binoculars and scan the sky more often.

Enthusiasm can be contagious. I hope so. If this book, which aims to record my enthusiasm and joy at searching the sky with my humble 'binoccies', helps you to catch the bug, then it will all have been worth the effort.

It needs to be emphasised that this book is intended as a guide and enthuser to those interested in studying the Southern Sky with binoculars. All the objects covered by this book are viewable in binoculars. Obviously, they are also viewable in telescopes but, in some cases, not as well because the extra magnification provided by telescopes can sometimes spoil the appreciation of the larger object.

Also, as the book's title implies, the objects are all viewable from the southern hemisphere. Other objects which are better (or only) viewed from the northern hemisphere are not included. There are books by others for these northern objects.

It is not intended to be a fully detailed technical reference for learning about the many and various aspects of astronomy. Occasionally, this book may touch on such details, but it is by no means an astronomy reference book.

Readers are encouraged to follow up such reference books to obtain a greater understanding of the history and science of what they will be seeing through their binoculars.

So, when by necessity of brevity I may use a technical phrase or jargon to describe an astronomical object, apart from a brief explanation in Part 4 of this book, I will assume you have access to a technical reference book for any further depth of information you may require.

The trick is to get out with your binoculars with the help of this book and be awed by what you'll see.

SECTION 2.

What's To Be Seen With Binoculars?
(What…I can see all that without a telescope?)

It's a very common misconception by people on the fringe of amateur astronomy that you absolutely need a telescope to "see anything interesting".

There is no doubt at all that to see the ultra-faint deep space objects (the faint fuzzies), you need the larger light collecting aperture of a telescope. Also, to resolve (that is, split apart) the closely spaced stars in multiple star systems and clusters, you need the magnifying power offered by a telescope.

No-one is denying that to continue the pursuit of amateur astronomy, delving deeper into the treasure chest of 'what is to be seen out there', a telescope is essential.

But when I learned to surf as a child, body surfing was more than enough fun. As the waves and I grew, a body board was the next logical step. After that would come a surf board. (It never did, by the way. I was happy body surfing.)

The point is, as a first step, and a very rewarding step it is too, binoculars have plenty to offer the beginner. After that, you will be in a position to decide if you want to take the next step to go deeper into the subject (and deeper into space) and purchase a small or medium sized telescope.

But even if you don't, with the help of this book (and no doubt others), binoculars will keep you occupied for many years.

When my astronomy society has an observing night under a dark sky, with ranks of telescopes from humble 63mm refractors (ex-chemist shop or supermarket) to the expensive 200mm or 250mm reflectors, I always take my binoculars and usually spend the first part of the long night using them before I start to peer through my telescope. I am rarely disappointed with what I see through the binoculars.

Having said that, it needs to be pointed out that the majority of sights described in this book do not require the perfect dark sky to be seen – just a reasonable cloudless sky from your suburban backyard.

Any description of what you can see with binoculars best follows a logical progression…from the closest to the furthest, from the brightest to the faintest. So, with your trusty binoculars and the help of this book, what will you be able to see?

Here's a brief summary. More detailed descriptions of each of these will be provided in the later sections.

The Moon.

It goes without saying that this clear, naked eye object (only 384,000 km away on average) is an easy and exciting target for binoculars. But you don't need me to tell you where to find it. However, you may be interested in a few facts about it to better appreciate what you are seeing. We sometimes take the Moon for granted because it's always there and always 'looks the same'.

The Planets.

Mercury, Venus, Mars, Jupiter and Saturn are the five 'naked eye' planets and each has something to offer binocular users. The next two outer planets, Uranus and Neptune, are also visible in binoculars but only after you know exactly where to look and you can pick them out from the background of stars. That is outside the scope of this book but still worth pursuing by other means.

Asteroids.

Some of these are visible to binoculars, but they will not be dealt with in this book as specialist information is required to know where to look, due to the peculiarities of their orbits. However, such information is readily available from the many mainstream astronomy magazines and their related websites.

The Stars.

The keen naked eye can only see stars up to 6th magnitude, but with the average pair of binoculars, you can see stars as faint as 9th magnitude.

A star seen through binoculars (or a telescope) is no different in appearance than when seen with the naked eye, except maybe brighter. Certainly not closer or bigger. However, binoculars do have the neat trick of making the star's natural colour more obvious. A red star looks more red in binoculars than it does to the naked eye.

To all intents and purposes, for an amateur telescope (as opposed to, say, the Hubble Space Telescope), a star is just a bright point of light.

But double, triple or multiple stars. Now that's a different matter. Binoculars can help 'split' what looks like a single star into two, or sometimes even three or more stars.

The Milky Way.

Our local galaxy, seen as a background of countless stars, is more easily appreciated through binoculars. The dense, rich fields of stars will leave you feeling very small. The words 'countless stars' take on a new meaning when viewing the Milky Way through binoculars.

Open Clusters.

There are many collections of stars which 'swarm' together in a loose, open pattern called an Open Cluster. More often that not, these are better appreciated through binoculars than a telescope, as a telescope, with its greater magnification, will bring them too close and you can't see the cluster for the stars.

Globular Clusters.
Whether sticking out like the proverbial 'sore thumb' or looking like a faint wispy ghost or a faint fuzzy star, these mini-galaxies are an exciting prey for cluster hunters. There are over 150 of these 'globs' surrounding our Milky Way, not all of them visible to amateur telescopes, let alone binoculars. But there are more than enough to work your way through with binoculars and, as indicated earlier, the Southern Sky has the best two, plus many more.

Comets.
When they come around, these are often best viewed in binoculars. Sometimes, too much magnification is not a good thing. Obviously, this book cannot describe the timing or location of particular comets as they come and go, at intervals of many years, or sometimes once only. However, when a comet, even a faint one, is in the sky, binoculars will give a great view. Information on individual comets is easily found in the mainstream astronomy magazines and their related websites.

Nebulae.
Faint, wispy clouds of gas, the birthplaces of stars and sometimes the legacy of the death throes of a star.

Galaxies.
You may be surprised to know that some galaxies are visible through binoculars. Admittedly they don't look anything like the beautiful coloured photographs you'll see in the books, posters and on the Internet. But they don't through a telescope either. You need large apertures and long time exposures to achieve that result. But these beautiful, awesomely remote islands of hundreds of billions of stars can be seen with humble binoculars. It's a sobering thought that just maybe, on a planet around a star in such a distant galaxy, someone may be watching the Milky Way through a pair of binoculars (or perhaps trinoculars?)

What You'll see.
We come to an important point. No doubt you've had occasion to browse through astronomy books, or seen photos in magazines or on the Internet, and the galaxies and deep space nebulae look absolutely beautiful. Rich detail of billions of stars, multi-coloured gas clouds, unique shapes, patterns and features.

The reality is that you do not see them like that, even when you look through a largish telescope. And certainly not through binoculars. If you were fortunate enough to be on an interstellar space ship and looked out a port hole at such a nebula or galaxy, it would not look like one of the photographs.

When you look through an eyepiece of any instrument, be it binoculars or a telescope, all you will see is what the retina of your eye can register at that moment. There will be no accumulated collection of light, unlike a time exposed photograph which builds the picture, photon by photon.

So be warned. When you look at these magnificent deep space objects through your binoculars, do not expect to see it as it looked in a reference book or magazine. The most vividly colourful nebula will still look like a wispy grey and white cloud, but the colour photos help to know what it does look like and it gives you a better appreciation of what you are actually seeing through the binoculars.

Johannes Kepler

One of the great pioneers of astronomy was Johannes Kepler (1571 – 1630). Coming out of an era when it was believed all the planets, and the Sun and Moon, orbited the Earth and into the Copernican heliocentric model (where all planets orbited the Sun), the model was still hopelessly inadequate as Copernicus had assumed all perfectly circular orbits. Kepler had access to detailed observations made by Tycho Brahe and after painstaking analysis (all by hand, no computers), he came up with his three famous laws of planetary motion.

1st Law: The planets move not in circles but in ellipses, with one focus of each ellipse being the Sun.
2nd Law: The radius from the Sun to the planet sweeps out equal areas in equal times.
3rd Law: The cubes of the average distances of the planets from the Sun are proportional to the squares of their orbital periods.

From Kepler to Newton

Isaac Newton (1642 – 1727) was born after Kepler died. Whereas Kepler had deduced his three laws from observation only, Newton, with his newly worked out laws of universal gravitation, was able to deduce theoretically Kepler's planetary model and laws which were confirmed by observation. Thus astronomers had a powerful theoretical tool to predict the orbits of moons and comets. In fact any celestial object which orbited another. Astronomy was on its way to becoming an exact science.

SECTION 3.

What Binoculars are Best?
(The Ins and Outs of Aperture and Magnification)

If you have already purchased your binoculars, then this section may be only of general interest in hindsight.

If you have yet to purchase them, then it may provide some basic hints as to what to look for – and avoid.

If at the end of reading this section you feel very confused about which apertures, magnifications, exit pupils etc you should buy and if you are fearful that an incorrect decision will ruin your stargazing enjoyment, take some heart from my personal experience. I was given my current pair of binoculars over 30 years ago as a gift. They are average to good quality, reasonably robust 12x50s. All that you will read below may suggest that they are not ideal for astronomy, especially for the fainter deep space objects. And technically that may be correct. However, I wouldn't swap them for the world (insult to the giver aside). They work for me and I have yet to be disappointed by them.

The point is – try to make a reasoned choice, then having made it, full steam ahead and damn the torpedoes.

Significant Features of Binoculars

The optics of binoculars are fairly complex and technical, involving many lenses and reflecting prisms. Like all quality items, 'you gets what you pays for'. My only word of advice on price is – don't buy cheap.

Suffice to say that they allow you to use both eyes at once (a very normal and comfortable practice for a human) rather than to squint one-eyed through a telescope. They also present you with an image right-way up (as opposed to a telescope which turns the object up-side down and left to right). This makes for easier moving about the sky.

The important technical features of a binocular can be reduced to: Aperture; magnification; exit pupil; field of view and; eye relief.

Usually, binoculars will have written on them at least two sets of numbers. For example, 7x50 and Field 7°. What this tells you is that the binocular has a magnification of 7 times, an aperture of 50mm and an actual field of view of 7°.

Let's look at the various features separately, then we can look at their overall impact on a choice for purchase.

Aperture:

The second number in the "7x50" (for example) written on the binoculars is the 'aperture'.

This is the diameter (in millimetres) of the main front lenses. This is where the light collecting happens. The larger the aperture, the more starlight that is collected and focused into your eyes, turning faint objects into brighter objects.

So, purely from an aperture point of view, the larger the better. (In astronomy, size does matter.) Obviously there are practical limits, such as cost, weight and steadiness of image.

For astronomy purposes, binoculars should have a minimum aperture of, say, 40mm. Smaller apertures may be great for the racetrack, bird watching etc, but they don't gather enough light for astronomy.

Binoculars tend to come in apertures of 42mm, 50mm, 56mm, 63mm, 70mm, 80mm and 100mm. (The latter three sizes are often referred to as giant models.)

The most common sizes used for amateur astronomy (both by those without telescope or even those with telescopes as well) are the 42mm and 50mm, usually the latter.

The 50mm is a good size, giving lots of light gathering, but not too heavy.

As a point of technical interest, the 50mm lens will gather 51 times the amount of light that the average human eye (with a maximum pupil size of 7mm) can gather. This explains why binoculars can see objects about 4 magnitudes fainter than the naked eye. (More about magnitudes of stars etc elsewhere in this book.)

Magnification:

This is both simple and a little tricky. It is simple in concept, but tricky when combining with Aperture and Exit Pupil (see below).

The first number in the '7x50' on the binoculars is the magnification. The magnification on a normal pair of binoculars is fixed. (No changing eyepieces like on a telescope to give different magnifications.) There are zoom (variable magnification) binoculars available but these are not recommended for astronomy as the optics can lack sharpness at the centre and edges of the image.

Magnification of 7 means it will appear to bring objects 7 times closer. Magnification of 10 brings them 10 times closer. This will make the object look 7 times (or 10 times) larger.

So a 7x50 pair of binoculars will make a small faint object (Orion Nebula, for example) 51 times brighter and 7 times closer (and larger) than seen with the naked eye. Or, a 10x63 pair will make objects 81 times brighter and 10 times closer.

7 times (or 10 times) larger may not sound a lot, but for astronomical objects, it is quite significant, especially in conjunction with the huge increase in brightness.

Now for the tricky bit. One may be tempted to get the best of both worlds in aperture and magnification (my own 12x50s are a good example). Be very careful, as there are downsides to high magnification. (We are talking here about different magnifications for the same aperture.) These are:

Wobble. If you are holding the binoculars by hand, with high magnification every little tremor of your hands will be magnified and you will have a very shaky image. To overcome this with large magnification, you had best invest in a binocular support stand.

Field of View. The higher the magnification, the narrower the field of view. This means you are seeing less of the sky in the eyepiece. This can sometimes make your object of study too large to see in one glance. It can also make you lose your orientation of where you are looking.

Brightness. If your magnification is too large (for a given aperture), it can lead to you losing the benefit of some of the light your aperture has gathered. This has the effect of your object appearing less bright against the background sky than it would with a lesser magnification. (This is often explained with an analogy of spreading the same piece of Vegemite over a larger piece of toast. The Vegemite is not as thick (bright) as it would be on a smaller piece of toast.)

Exit Pupil:

The explanation for the loss of brightness with higher magnification is simple. It has to do with the size of the image exiting the binocular's eyepiece. If you divide the aperture (say 50mm) by the magnification (say 7), this gives an exiting image diameter (called the Exit Pupil of the binoculars) of 7.1mm which, happily, is the size of the average eye's maximum pupil size (7mm to be exact). That is, the full image is able to fill the eye's pupil. There is no loss of image nor wasted pupil area. The ideal situation.

But suppose the binoculars were 10x50. This gives an Exit Pupil of only 5mm diameter. This means the pupil, open to 7mm, is receiving only 5mm of image. This gives a 50% reduction of image brightness, as only half of the eye's light gathering capability is being used.

Therefore, **as a general rule**, all binoculars with a 7mm exit pupil (say 7x50, 8x56, 9x63, 10x70) are the best for astronomy, at least where one wishes to view fainter objects.

BUT... I still revel in my 12x50s and others are happy to use 10x50s. Why?

It's sad but true that as you get older, your maximum pupil size reduces. This varies with individuals, of course, but again as general rule, over 30s have a maximum pupil size of 6mm and over 40s have about 5mm to 4.5mm.

So, if you are older and have a maximum pupil size of say, 5mm and you used 7x50 binoculars with an exit pupil of 7mm, the light in the outer 1mm wide ring of the image (50% Of the light) would not enter your eye and would be wasted. But if you used 10x50s, the exit pupil of 5mm would exactly match your maximum pupil size and your eye would see the full image with the optimum maximum brightness.

My 12x50s (an irregular magnification, I might add) give an exit pupil of 4.1 mm which is probably a little small for my eyes with 5mm maximum pupil. However, I don't mind the marginal loss of brightness in order to gain the slightly larger magnification.

However, if I were ever (sadly) to lose my 12x50s, I would probably go out and buy a pair of 10x50s (or 16x80s if I felt adventurous).

Field of View:

Astronomers measure the sky in angles, or degrees. For example, the Moon, as seen from Earth is half a degree (0.5°) diameter while the Southern Cross is about 4° wide.

Each binocular will have written on it the Field of View (say 7°) which is the diameter (in degrees) of that circle of the sky which you can see through the binocular. This is a consequence of the optics of the binoculars, but in general it tends to reduce as the magnification increases.

For general astronomy viewing, a field of view (FoV) of 7° is considered satisfactory. Some 10x50s may have a FoV of about 5°.

Eye Relief:

Eye relief is a measure of how far you need to place your eyes from the surface of the eyepiece to receive the whole field of view.

This is a personal comfort factor. You don't want to have to press your eyes hard up against the eyepiece to receive the exit image. It is both uncomfortable and can result in smudging the eyepiece with eyelash dust etc. This would be the outcome of binoculars with eye reliefs of 9mm or less.

Larger eye reliefs (say about 14mm to 15mm) are both comfortable for normal eyes and also practical for people who wear glasses for near sightedness. Rather than adjust the binocular's focus to compensate for your vision, you can look at the sky with your glasses, locate what you want to view, then move the binoculars in front of you with your glasses still on. This is much easier.

Some Other Factors:

Binoculars should be robust to survive the hurly burly of field work in the dark. Look carefully at the material and coating of the binoculars. Give preference (if you can afford it) to binoculars whose body has a hard rubber coating and rubber moulds around the glass surfaces. The latter provides protection against those inevitable knocks and drops on the ground.

Of course you will need a sturdy case with flip lid to carry them in. This protects them from damage during transport.

And the binoculars should have a strong cord for hanging around your neck. NEVER use the binoculars without first hanging the strap around your neck. Otherwise you are courting disaster. Besides, it gives you somewhere handy to put them when you want to use your hands for other things, such as pouring a hot coffee from the thermos on a cold night.

To Sum It Up

As I said earlier, it's a case of horses for courses. You need to settle on an aperture size (at least 42mm, preferably 50mm) that is both comfortable to hold (bigger means heavier) without significant wobble and within your budget. Then decide on what magnification you want. To do that, you'll need to take into account the issues of exit pupil size, your own maximum pupil size, and what field of view you'd be happy with.

Be prepared to compromise. Decide what sacrifices you are prepared to make in terms of image size (magnification) versus brightness. For the brighter objects such as multiple stars and clusters, magnification may be more important than retaining optimum brightness. (10x50 may then be appropriate.) On the other hand, for the fainter objects such as nebulae and galaxies, retaining optimum brightness may be preferred to making them look that little bit bigger. (The 7x50 may be better.)

Also, it should be remembered that you may wish to share the binoculars with family members, all with different maximum pupil sizes. More compromises required.

Personally, I prefer that little bit of extra magnification from a 10x50, even if I had perfect eyesight and had to sacrifice a little brightness.

But, it's finally up to you. Whatever you choose, I'm sure you'll get great enjoyment from your binoculars.

Splitting Double and Binary Stars

A person with 20/20 vision should be able to split two stars that are four arc-minutes (4') apart in the sky. The average eye cannot resolve anything closer than that. That means the naked eye can resolve down to about $1/7^{th}$ of a Moon diameter. Try and picture that.

There are 60 arc-minutes (60') in 1 degree (1°), and there are 60 arc-seconds (60") in 1 arc-minute (1'). So an average naked eye should be able to split anything 4' (240") or more apart.

This means that when you look at a double star in binoculars, it magnifies the _actual_ angular separation of the two stars by the magnification of the binoculars. What your eyes see is the magnified separation. If _that_ is less than 4', then you shouldn't be able to split them. If it is greater than 4', then you should.

However, there are always other factors, the biggest being a significant difference in the magnitudes of the two stars. If one is much brighter than the other, even if wider than the minimum separation, the fainter one can be lost in the glow of the brighter. Then, of course, the 'seeing' conditions on the night, the steadiness of the binoculars, your own eyesight, can all have an impact.

In general, except in extreme differences in magnitude, any separation that magnifies to greater than 10' should be able to be split.

SECTION 4.

I See It, But What Is It?

A great part of the excitement of astronomy, whether using binoculars or a telescope, is appreciating just what it is you are seeing. This gives you the *"Wow!"* factor.

As an example, if I pointed out to you the red star Betelgeuse in the constellation Orion, you would probably say "Nice red star, so what?"

Then if I explained that Betelgeuse is one of the largest known red supergiants with a diameter over 400 times that of our Sun (that is, over 560 million km) and a brightness of 14,000 of our Suns; if it was placed at the centre of our Solar System it would swallow up Mercury, Venus, Earth, Mars and a bit more; and that its material is so thin, its mean density is less than one ten-thousandth of that of the air we breath, so thin that it is often described as a "red hot vacuum"; and that it is a prime candidate to go supernova very soon, leaving a black hole only 470 light years from us; then you would look at this bright red bauble in the sky and say **"Wow!"**

They say "seeing is believing". True. But "seeing and knowing" is inspiring.

This section of the book is intended to provide the "knowing" half of the **Wow!** factor. It will give some background details of the various types of objects described in the "What's to be seen with binoculars" section. You won't need a science degree to understand the information. Just some imagination and, in most cases, a willingness to think big. Space is a very big place, as the next section explains.

The Scale of Space
(or Big just doesn't begin to describe it).

Even to the uninitiated Space is obviously very big. But it's not until someone starts spouting numbers that you begin to realise just how big.

It will help to appreciate what your binoculars are showing you and certainly add to the **Wow!** factor if the various items of information provided in this book of the distances to planets, stars, clusters and nebulae, not to mention the galaxies are understood. Even then, to understand a staggering distance is one thing, but to be able to imagine or comprehend it is quite another. Our human minds can only grasp so much. Let me use the following model to put it in perspective:

Let's assume that our Sun is the size of a small grape, only 1 cm in diameter, in the centre of the cricket pitch at Sydney Cricket Ground.

Where would the Earth be? It would be the size of a grain of sand, orbiting 1 metre from the grape (the Sun). That's not so hard. You can imagine that.

Where would Saturn be? It would one twelfth the size of the grape, orbiting around the stumps at each end of the cricket pitch.

And what about Pluto, the edge of our solar system? It would be at about the start of the fast bowler's run up, half-way to the boundary rope.

That's OK so far. We can imagine that... just.

So how far is it to the next closest star, the binary Alpha Centauri, a pair of similar grapes? It will be 295km away, the distance from Sydney to Bateman's Bay on the South Coast. And there is nothing – absolutely nothing - in between.

But if you want BIG, how far is our grape-sized Sun from the next nearest major galaxy, the Andromeda Galaxy? On this scale, it is slightly more than the distance from the real Earth to our real Sun – about 162 million km. And on the same scale, the Andromeda Galaxy, like our own Milky Way, would be 10 million km in diameter.

Now that was if our Sun was the size of a 1 cm grape. But the Sun is actually 1.4 million km in diameter, so all the distances in our little grape model have to be multiplied by 140 billion.

Try and imagine that, and that's only to the closest galaxy. It is currently believed that the distance to the furthest observable galaxies (by huge professional telescopes) is some 5,000 times further than the Andromeda Galaxy. So when people say the universe is unimaginably big, they are not exaggerating. You can quote the numbers, but you simply cannot *imagine* them.

Getting back to the numbers:

When we describe distances within our solar system, we usually use the conventional kilometre (km), though as we go out towards its outer reaches, it's not unusual to employ a larger unit called the Astronomical Unit.

The Astronomical Unit is simply the average distance from the Earth to the Sun which is 149,492,000 km (to the nearest thousand). So the distance from Earth to another planet, or the distance of that planet from the Sun can be stated in millions of kms, or in Astronomical Units (AUs). For example, the mean distance of Mars from the Sun is 228 million km, or 1.524 AUs. This idea becomes more useful when you think about the furthest planets like Uranus, Neptune and Pluto which have mean distances to the Sun of 19.2, 30.7 and 39.7 AUs respectively. You can picture that Pluto is about 40 times further from the Sun than Earth is.

But when we start to talk about distances to the stars, kms and AUs just don't cut it. There are too many of them to handle. For example, the distance to the nearest star, Alpha Centauri, is 41.3×10^{12} km (41.3 thousand billion km) or 276,560 AUs. And the distance to the Andromeda Galaxy? 23×10^{18} km (23 billion billion km) or 152,000 million AUs. So a bigger unit of distance is needed to make such numbers manageable.

Enter the **Light Year**. The light year is a unit of distance, not time, though ironically, it has an equivalent use of telling you how far back in time you are looking. More of that later.

One light year is simply the distance that light will travel through a vacuum in one year. As the speed of light (in a vacuum) is 300,000 km/second, this gives a distance of 9.46 x 10^{12} km (or 9.46 thousand billion km or 9.46 trillion km).

So the distance to Alpha Centauri can be stated as 4.37 light years and the distance to the Andromeda Galaxy is 2.4 million light years. The distances are just as big but the numbers are much easier to grasp. The danger is it can lull you into a sense of complacency. You can imagine the number, so you think you can imagine the distance.

To shake you out of that fantasy, remember that the light year distance measure also gives you an indication of how long ago the light you see left the object. This is an exciting thing about astronomy. Your binoculars are effectively part of a Time Machine. Everywhere you look into space, even to the humble nearby Moon, you are looking back into time. You *never* see anything exactly as it is now, only as it was when the light you see now left it.

So when we look at Alpha Centauri, we see it as it was 4.37 years ago. When you look at the Orion Nebula, you see it as it was 1,500 years ago. And when you look at the Andromeda Galaxy, you see it as it was 2.4 million years ago. You are seeing the past, not the present. That is a fantastic concept and I still can't quite get used to it.

So if you start to think a certain star cluster is not so far away because it is 'only' 2,000 light years away, just remember that the light you see took 2,000 years to reach your eye. That makes it seem a lot further away.

Now back to the details of what your binoculars can show you. Obviously, for reasons of space (no pun intended), the details will need to be kept brief. After all, there are whole volumes out there written about each topic and new discoveries are constantly being made.

The Moon:

Although the Moon always shows us the same face, its appearance constantly changes due to its cycle of phases. These phases result from the direction the Sun is shining on it, which in turn results from the position of the Moon in its orbit about Earth. This is shown in the next diagram. The outer circle shows the position of the Moon in relation to the Sun and the Earth. The inner circle shows the resulting view of the Moon's phase seen from Earth.

The different phases present different views of features of the Moon due to the directions of shadows of craters and mountains.

Believe it or not, a Full Moon is not necessarily the best time to study the Moon because the lack of shadows gives it a 'washed out' or featureless look.

Try watching the Moon over changing phases and note the differences. It takes 29.5 days for the Moon to go through a complete cycle of phases, while it actually orbits the Earth in only 27.3 days, the same time it takes to spin once on its axis. (This is why it always offers the same face to Earth.) The difference of 2.2 days is a result of the Earth moving in its orbit about the Sun, changing the angles of sunlight that cause the different phases.

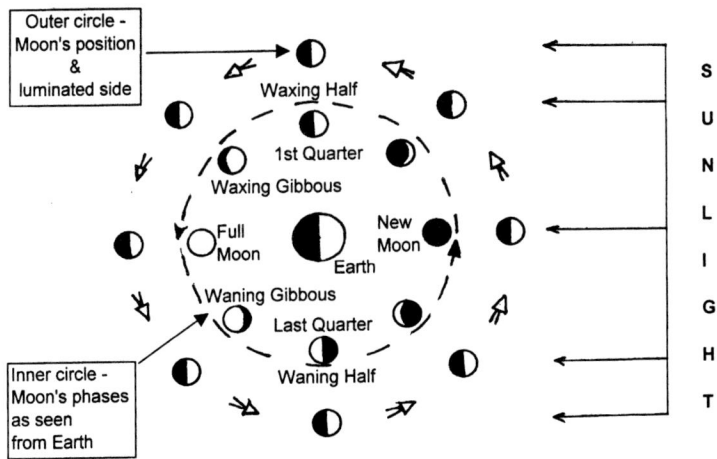

The line on the Moon's surface where the light meets the dark is called the Terminator. This moves across the Moon's surface as the phases gradually change. It's at the Terminator that the contrasts created by shadows are the greatest and the craters and mountains appear most dramatic.

So, if you want to study the Moon, try and pick those nights when there is a 'non-full' Moon. This may be either at First or Last Quarter, a crescent Moon (between New Moon and First or Last Quarter) or a gibbous Moon (between Full Moon and First or Last Quarter). If you can steady your binoculars, you should be able to identify the major features of 'mares' and craters by comparison to a reference map.

A few facts:

- The Moon's diameter is 3,476 km, about one quarter of Earth's. Its gravity is about one sixth that on Earth.
- Its average distance from Earth is 384,000 km. Since light travels at 300,000 km/second, it means that the light from the Moon took 1.28 seconds to reach your eye. (That is the first step in what I call 'the time machine of astronomy'.)
- There is no atmosphere on the Moon and each location on the Moon receives alternately two weeks of sunlight, then two weeks of shade. As a result, its surface temperature heats up to 100°C (the boiling point of water), then cools down to a chilling −170°C.
- The so-called 'seas' on the Moon (for example Mare Imbrium) are of course not seas at all as there is no water on the Moon. (At least not in liquid form. There was a recent discovery of frozen ice at the bottom of a deep crater on the far side, supposedly enough to supply a colony. But it's the only 'water' discovered so far.) The seas are actually large areas which had been flooded by dark volcanic lava eons ago. But these are not perfectly smooth, having

been pockmarked by ages of meteorite bombardment.
- All the craters visible in binoculars (and those too small to be seen in binoculars) are the result of constant bombardment by meteorites, large and small, over the billions years since the Moon's volcanic activity ceased and its surface cooled. If it were not for the Earth's protective atmosphere, we would have suffered the same fate.
- All the major craters, mares and mountain ranges have names. Detailed Moon maps are available at astronomy book shops which clearly identify each feature. There are many amateur astronomers who enjoy picking out individual craters etc and locating their names from the maps.

The Planets:

There are five 'naked eye' planets which are also rewarding to view through binoculars. These are Mercury, Venus, Mars, Jupiter and Saturn. While all five are worthy of attention, without a doubt, Jupiter is the most fascinating and likely to draw you back to it again and again.

The word 'planet' comes from a Greek word for 'wanderer'. This is because they were seen to wander across the sky against the background of stars due to the combined effect of their's and Earth's orbit about the Sun.

For that reason, this book cannot specify for any particular date where you will be able to look for any particular planet. But there are some simple solutions to this.

Firstly, the Weather Section in most daily newspapers will have a table giving the rise and set times for that day of the Moon and the five naked eye planets. From this you can calculate the approximate distance from the eastern horizon that the planets will appear that night (if they will appear at all). Depending on where the planet is in its orbit about the Sun, there are months when the planet is simply not visible as it is 'up' during daylight.

The calculation is simple. Because of Earth's rotation (West to East), the planets climb from the Eastern horizon towards the West at a rate of $15°$ per hour. So, for example, if Jupiter were to rise (appear on the Eastern horizon) at 6pm, then at 8pm it will be $30°$ from the Eastern horizon and by midnight it will be $6 \times 15° = 90°$ from the Eastern horizon, approximately due north.

Secondly, there is a better way to predict the observation times of the planets, especially if you want to plan ahead by a few days or weeks. You can buy an ephemeris, at a very reasonable price, which sets out for the whole year, month by month, all the locations of the planets with their rising and setting times. Amateur astronomers totally depend on their ephemeris.

One excellent ephemeris, suitable for eastern Australia is called 'Astronomy 200X", where the 200X is the year to which it applies (e.g. 2004). It is written by Dawes, Northfield and Wallace and put out by Quasar Publishing and is available at any astronomy supply shop. It contains all the information about the solar system that any amateur astronomer – beginner or experienced – could hope for, and more.

With this, you can look ahead and see when a particular planet will be available for viewing at a 'civilised' hour, how bright it will be, what its moons are doing, and what constellation you will find it in.

It also gives details about future conjunctions (when two celestial objects come very close together – usually forming very attractive arrangements in the sky) and occultations (when one celestial object passes in front of another hiding it from view) of the planets, usually with each other and the Moon, but also sometimes with bright background stars.

When searching the sky for the planets, remember they don't go directly from East to West. They follow the imaginary curved line in the sky called the 'ecliptic'. The ecliptic is the path the Sun follows in the sky as it travels from East to West. This is a result of the Earth's axis being tilted 23.5° off the vertical, so the imaginary plane through the Earth's equator is at an angle of 23.5° off the plane of Earth's orbit around the Sun. Depending on where you are and the time of year, it will tilt away from the East-West line either towards the North or the South.

Now the naked eye planets generally follow this line in our sky as well (if not right on it, then very close to it), as their orbits are in the same plane (give or take a few degrees) around the Sun as Earth's. So if you know the path the Sun follows in the sky at your location for that time of year, then that is the path the planets will be found travelling from planet-rise to planet-set.

At certain times of some years, all five naked eye planets are 'up' on the same night, and it is a delight to pick them out on the ecliptic as you sweep from East to West, like a line of straggling sheep.

Now, let's have a closer look at each of the five naked eye planets.

Mercury:

Mercury is the closest planet to the Sun. Because of this, it is not the easiest to observe as it is only visible close to the horizon just before sunrise or just after sunset, mostly during twilight when it is not fully dark.

However, at certain times of the year, it reaches what is called 'maximum elongation' from the Sun and stays above the dark horizon for a short while, up to half an hour.

You will need an uncluttered western or eastern horizon to see Mercury. It doesn't appear high above neighbourhood trees or houses. This is what makes it more of a challenge to catch a glimpse of the planet.

To be realistic, a glimpse is about all you can hope for, even with binoculars or a telescope. Because of its small size, proximity to the Sun and lack of distinctive features, Mercury in binoculars (and larger telescopes) will mostly appear as a bland orange blob. Its magnitude during the year can range between −2 and +1.

Mercury does go through phases (like the Moon) during its orbit about the Sun. However, as it is never bigger than 11" in angle, binoculars cannot show them to you.

A Few Facts:
- Mercury is named after the fleet footed messenger of the Roman gods.
- It orbits the Sun in a mere 88 days.
- It has a diameter of 4,880 km, only 50% larger than our Moon, or 38% of the Earth's diameter. It is more appropriate to compare it to the Moon, however, because that is exactly what it looks like. If you saw photos of our Moon and Mercury taken from space probes, you'd barely be able to tell them apart. Like our Moon, Mercury is covered with craters, large and small.
- Mercury has some paradoxical aspects of its day and year. Its year is 88 Earth-days. However, Mercury rotates on its axis once every 59 days. This rotation, combined with its solar orbit, results in the time between two consecutive sunrises on Mercury (a Mercury day) taking 176 Earth-days. That is 2 Mercury years. That's really weird when you think about it.
- Because of the long day, plus the proximity to the Sun which is a mean distance of 58 million km, the temperature on the day side of Mercury soars to 400°C, while the night temperature plummets to –170°C.

Venus:

The second planet from the Sun, Venus is probably responsible for more UFO sightings than any other cause. Named after the Roman Goddess of Love, Venus is often identified by non-astronomers as the Morning or Evening Star. It has conjured up in novelists' minds images of a sister planet, suitable for habitation. A veritable paradise. In fact, nothing could be further from the truth. It is a living hell.

Like Mercury, Venus is most often seen at twilight (morning or evening) hovering above the horizon like a signal flare or a Jumbo jet coming in to land. Hence the UFO sightings. However, there are times of the year when Venus remains quite high in the sky long after sunset. This often confuses people who only think of Venus as a horizon object.

Venus can be incredibly bright, reaching a maximum magnitude of –4.7, which is about 7 times brighter than Jupiter's maximum brightness. The reason for this brightness is mainly due to the very clouds that cause its hellish environment, rather than its close proximity or size as commonly thought. The clouds, which totally conceal Venus' tortured surface, reflect 75% of the Sun's light compared to Mercury's cratered surface which only reflects 10% of the Sun's light. Venus is an extreme example of the Hothouse Effect.

The combination of the carbon dioxide in the atmosphere and the sulphuric acid (yes, 80% concentration sulphuric acid) and water vapour in the clouds result in the retention of most of the heat transmitted from the Sun through the clouds. Just like a garden hothouse, only much more effectively.

Probes that have descended through Venus' atmosphere and survived to reach the surface have registered temperatures of over 470°C, which is hotter than molten lead. Then the probes collapsed, both from the heat and the atmospheric pressure that was over 90 times that of Earth's.

As a final touch, the clouds cause a constant drizzle of rain to fall – sulphuric acid more corrosive than the strongest battery acid.

Hell indeed!

Hidden beneath this malevolent atmosphere is a landscape of rugged beauty, though no human eye will ever see it directly. We know about it from radar scans from space probes which have surveyed the planet's entire surface. The landscape contains mountain ranges higher than Mt Everest, craters up to 100 km wide and huge valleys running like jagged wounds thousands of kilometres long. There are also active volcanoes, with evidence of fresh lava flows.

All this is hidden behind an apparently benign, featureless milky white planet, as it will appear in binoculars. Because when Venus is conveniently close, it will show up in binoculars as a distinct, white, flat disc. With very distinctive phases.

That's one of the fascinations with Venus, watching its phases, just like our Moon's.

Paradoxically, we see Venus at its brightest (magnitude –4.7) when it is in a very crescent phase stage, showing only 28% of its disc to Earth. This is because the particular combination of its distance from Earth at that phase gives the greatest brightness. When Venus is on the other side of the Sun from Earth, it will give a fuller phase and a higher percentage of its disk, but it is much further away and therefore less bright.

So, when you are observing Venus with your binoculars, admiring its phase and apparent white milky purity, remember the boiling hell that lies beneath it.

A Few Facts:

- Venus has a diameter of 12,104 km, only 625 km less than Earth's. It is virtually a twin to Earth in size.
- Its mean distance from the Sun is 108 million km.
- Its time to orbit the Sun (a Venus year) is 225 days (Earth days).
- Its closest approach to Earth is 40 million km.
- Venus rotates about its axis in the opposite direction to all the other planets, including Earth. It rotates from East to West.
- Venus rotates very slowly, once in 243 days. In fact it completes an orbit around the Sun (a Venus year) before it completes one of its rotations.
- Due to its hothouse atmosphere, Venus has a constant temperature at all points on its surface in excess of 470°C.

Mars:

There used to be a time when astronomers had to wait every 15 years to get their best look at Mars during times of opposition. That was because with Mars' very elliptical orbit and Earth's near circular orbit, Mars was at its closest to Earth (55 million km) only once every 15 years. At its furthest opposition with Earth (100 million km) it was very difficult to view in any detail.

Of course this has all changed with the Mars space probes in the past years (e.g. the Pathfinder Global Surveyor and Exploration Rovers Spirit and Opportunity) with photographic surveys of the entire planet from close orbital range.

Alas, a lot of the mystery has disappeared with scientific precision stepping in.

But for us mere mortals with binoculars, the mystery can remain. While still only appearing as a non-twinkling orange star, it's still the Red Planet, named after the God of War (and Venus' lover), the launching pad of countless B-Grade Earth-invading aliens. Mars' warlike character was perfectly captured in the magnificent music of Gustav Holtz's "The Planets" suite.

We can thank the American astronomer Percival Lowell for most of that, after he mistranslated Giovanni Schiaparelli's 'canali' (meaning channels) for the word 'canals', conjuring up images of artificial commerce carrying water ways.

The truth, sadly, is much less romantic. Mars is (we believe until proven otherwise) a lifeless planet, at least now, if not millions of years ago. Its atmosphere is unbreathable carbon dioxide with no oxygen, and as thin on the surface as our atmosphere is 30 km above ground. Its climate has extreme temperatures and apart from ice on its small Polar Caps and an inaccessible sublayer of permafrost, there is no water on Mars in a practical sense.

Having said all that, Mars is not without its unique features and the odd **Wow!** factor.

For example, Mars is the home of the largest known volcano in the solar system, Olympus Mons. It is 26 km high and has a massive 600 km wide base. Its sloping sides are so gentle and its 'peak' so far away that if an explorer (in a space suit of course) was deposited half way up its slope, he'd have trouble figuring which way to walk to go 'up'.

Another extreme feature is an unimaginably deep, wide and long rift in the surface, called the Valles Marineris. It is up to 4 km deep and 600 km wide in places and 4000km long. That is, longer than the distance between Hobart and the tip of Cape York. This is obviously one of the features seen by Schiaparelli and labelled 'canali', which started all the canals fuss.

But despite the hard scientific data available about Mars, there will always be that element of mystery and romance. People want to believe there was once life there. Hence the brouhaha about the Face on Mars, an apparent monument with features of a human face, captured by an early space probe. Proof of an ancient civilisation, no doubt! Unfortunately, more recent surveys have revealed it to be what more rational people always knew. A random arrangement of hills and craters on a large mesa, casting a coincidentally familiar shadow at a certain angle of sunlight.

So when viewing Mars, admiring its ruddy red hue, imagine what might have been millions of years ago. And simply enjoy.

A Few Facts:
- Mars has a diameter of 6,790 km, a bit over half of Earth's diameter.
- Its gravity is 0.38 of Earth's. That's about twice the gravity on our Moon.
- Mars has two small moons, Deimos and Phobos. These are not visible in binoculars.
- Its mean distance from the Sun is 228 million km, with a maximum of 249 million km and a minimum of 206 million km.
- Mars' year is 687 Earth days, while its day is 24 hours, 37 minutes.
- At its brightest, Mars shines at mag. −2.8 and can be as faint as mag. +2.
- Mars' surface temperature ranges from a summer afternoon of −29°C to a chilly −100°C in winter.
- Despite the extremely thin atmosphere, there are severe storms on Mars, with winds in excess of 200 km/hour. These blow up large amounts of dust which can hide huge areas of surface features from telescope view.

Jupiter:

Jupiter is the fifth planet from the Sun and a rewarding and fascinating object to view through binoculars. If Jupiter is in the night sky, it is very difficult to miss. Even with its huge distance from Earth, ranging from 509 million km (at its closest) to 960 million km (at its furthest), it is brighter than every star except Sirius. The only other object to outshine it (other than the Moon) is Venus.

The reason for Jupiter's brightness is two-fold. Firstly, Jupiter is not a solid planet like Earth, Mercury, Venus and Mars, but a gas giant. It is mostly comprised of hydrogen and helium gas, like the Sun.

It is totally covered by swirling clouds of ammonia, and these clouds reflect half (50%) of the Sun's light landing on Jupiter. (Compare this to Mars which reflects only 15% of the sunlight, or Mercury which reflects 10%.)

Secondly, Jupiter is big, as in B-I-G!! The dimensions of this planet stagger your imagination. It is approximately one tenth the diameter of the Sun. Its diameter is eleven times that of Earth. Put another way, you could fit 1300 Earths inside the volume of Jupiter.

If you added up the mass of all eight of the planets without Jupiter, then Jupiter's mass would be 2.5 times that total. That is, Jupiter accounts for 71% of the mass of all nine planets.

However you look at it, Jupiter is a veritable giant – a King of the Planets.

Jupiter's surface is a very complex thing, as the various chemicals in the atmosphere are constantly rising, descending, swirling and curling. There have been measurements by probes of winds up to 600 km/hour.

But it's what's below this swirling colourful surface that is so fascinating. First, there are complex hydrogen based chemicals that would make an industrial chemist envious. It's this mix of chemicals that gives Jupiter its kaleidoscopic appearance. Then beneath these chemicals is a thick layer of water vapour just like Earth's clouds.

But below these it gets weird. Because of the weight of over 1000 km of ammonia, chemicals and water vapour, the pressure builds up to immense levels. The crushing pressure at the bottom of Earth's oceans is nothing compared to this. So crushing is the pressure that the hydrogen becomes a liquid and there is an 'ocean' shell of liquid hydrogen about 20,000 km deep.

And below this ocean layer? The liquid hydrogen ocean adds again to the pressure, so much so that that the hydrogen 'gas' below it is compressed into a form so dense it is virtually a metal, something like the metallic mercury in a thermometer. It is this metallic hydrogen that leads to the huge magnetic fields generated by Jupiter.

And below that metallic hydrogen…who knows? Perhaps a core of rock, as some think. No-one will ever know.

Wow?

Meanwhile, back on the surface, Jupiter presents a face covered with multicoloured bands of moving gas. Each band is moving at different speeds, swirling and interacting at the borders. Jupiter's face never stays the same. It also has the famous Great Red Spot, which is believed to be a hurricane type storm that has lasted on Jupiter for at least 400 years. It was there when the first telescopes were aimed at Jupiter in the early 1600s, so it's at least that old. This 'storm' is so large that you could fit three Earths along it. Batten the hatches!

All this gives Jupiter the appearance in a telescope of a very colourful lolly pop. You will not see all this detail in binoculars, however, as magnification larger than the 7x or 10x available from binoculars is needed.

What will you see of Jupiter in binoculars?

Firstly, you will see Jupiter's disk as a pale creamy plate. It will be a solid disk, not a large twinkling star. You will be seeing Jupiter's actual surface, even though indistinctly. Depending on Jupiter's distance and the size of your binoculars (here's where magnification counts), that creamy disk may be quite prominent. It is unfortunate that the optics of binoculars is just below the level of showing the cloud bands and the Red Spot, which even a small telescope will show.

But it is still fascinating to see this disk and know that you are actually seeing the surface of this Giant of Planets, over 600 million km away.

Secondly, and this is what places Jupiter in the World's Hall of Scientific Fame, you can see the four main moons of Jupiter. These have been dubbed the Galilean Moons, in honour of Galileo Galilei who saw them in 1609 A.D. when he first turned his puny telescope towards the heavens. When he realised that Jupiter had its own system of moons which orbited around Jupiter (and not around Earth) it gave credence to the Capernican theory that the Earth and the other planets orbited the Sun, rather than everything orbiting the Earth.

Putting aside the consequent problems with the Church at that time, this discovery of the moons of Jupiter was a momentous event in the history of astronomy and science.

And you can see them with your binoculars, probably more clearly than Galileo did with his crude telescope.

The Galilean moons will appear as tiny specks of light in line abreast of Jupiter's disk. And what is more fascinating is that you can watch them move from night to night.

The four moons, in order of distance from Jupiter are: Io (the closest), Europa, Ganymede and Callisto. With the exception of Europa, these moons are larger than Earth's moon.

Io has a diameter of 3,660 km and a period of orbit of 42.5 hours (1.77 days).

Europa, with a diameter of 3,130 km, has a period of orbit of 85.2 hours (3.55 days).

Next is Ganymede, the largest moon in the entire Solar System, with a diameter of 5,258 km. Believe it or not, Ganymede, a mere moon of Jupiter, is even larger than the planet Mercury and more than double the size of Pluto. It has a period of orbit of 7.15 days.

Finally, Callisto, also larger than our Moon (but not quite Mercury) at 4,806 km diameter, has an orbital period of 16.7 days.

All four Galilean moons are orbiting at different distances out from Jupiter, but they are all in the same plane. This gives them the appearance from Earth of all being in a straight line. This is not strictly correct as their plane of orbit is slightly tilted to Earth, but to all intents and purposes, especially with binoculars, they appear to be in line. And because of their different periods of orbit, we end up with the famous 'Dance of the Moons.'

Galilean moons' positions over a typical set of 10 nights

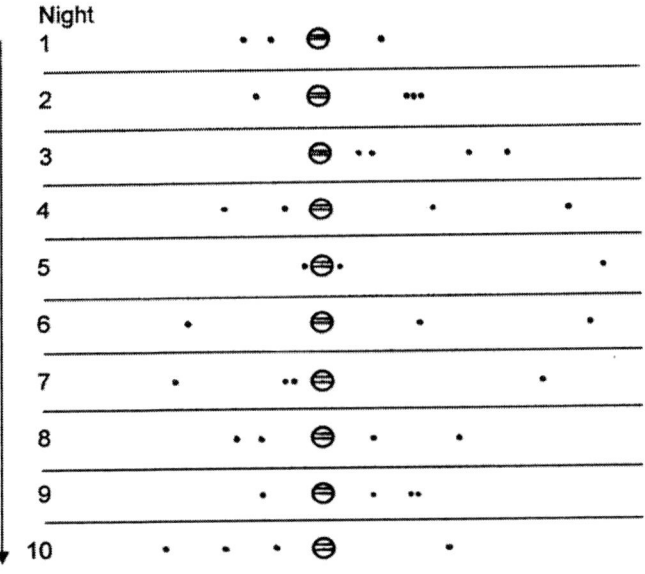

It is fun to watch them move about from night to night. One night there are two moons on each side of Jupiter, the next night there are three and one. The big event is when you see all four moons lined up on the same side of Jupiter. Another special treat is when you watch one moon actually pass behind another. Two pin-point dots of light slowly come closer and then become one. After a brief period, they separate again into two. The reality is that one of the two is actually hundreds of thousands of kilometers behind the other in its orbit, but they look like they are colliding.

And when one or more of the moons is 'missing', then you know that they are either passing in front of or behind Jupiter. They will 'pop out' eventually. These Moon Dance events have been precisely predicted and you can plan your viewing ahead, either by reference to *Astronomy 200X* or the mainstream astronomy magazines. You will get the best view of the 'dance' if you can steady your binoculars to minimize wobble.

But always remember, when you use your binoculars to spot Jupiter's moons, you are observing *exactly* the same thing that Galileo did and turned astronomy on its head.

A Few Facts:

- Jupiter has a diameter of 142,984 km, approximately one tenth the diameter of the Sun.
- Its gravity is 2.64 times that on Earth.
- Its mean distance from the Sun is 777.8 million km (5.2 AUs), with a minimum of 740.9 million km and a maximum of 815.7 million km.
- It orbits the Sun once every 11.9 years.
- Jupiter's "day" is a mere 9 hours 50 minutes. This is amazing when you consider its size (about 11 times the diameter of Earth). In fact this is the fastest rotation speed of all the planets.
- At maximum brightness, it reaches mag. –2.9.

Saturn:

Saturn is the sixth planet from the Sun and is famous for its rings. The ordinary person in the street who knows absolutely nothing about astronomy knows about Saturn and its rings.

Saturn presents itself to the naked eye as a yellowish 'star' of magnitude up to –0.3. This is basically due to its huge distance away from Earth. Saturn has an average distance from the Sun of 1,426 million km or about 9.5 AUs.

However, if it wasn't for its famous ring system, Saturn would be a lot fainter. The fact is that the rings reflect more sunlight than the planet itself and when Saturn is at its brightest, more than half of that is because of its rings.

It should be noted that because the plane of Saturn's rings is not exactly the same as the plane of Saturn's orbit about the Sun (Saturn and its rings are tipped 29° from the vertical), there are times when the rings are edge-on to our view and seem to disappear completely. This happens about every 15 years, being half of Saturn's orbital period around the Sun.

Like Jupiter, Saturn is a gas giant. At 120,536 km diameter, it is slightly smaller than Jupiter, but surprisingly it is only about one third of Jupiter's mass. Overall, its chemical make-up is very similar to Jupiter's, but for some reason it does not have the multi-coloured cloud bands and Big Red Spot that Jupiter has. Perhaps this is nature's compensation for the glorious rings.

The rings, which look like a giant pancake, have an overall diameter of 270,000 km. But they are unimaginably thin, with a maximum thickness of only 100 metres. If you had a pancake 5mm thick, it would have to be 14 km in diameter to be on the same scale as Saturn's rings. (A definite **Wow!** Factor.)

Astronomers are still debating exactly how the rings formed, whether they are made up of planetary disk material that failed to lump together as a moon, or whether they are in fact the remnant of a moon that actually disintegrated for some reason. It is generally agreed that the rings' material is mostly lumps of water ice and dust, with sizes ranging from a speck of sand to a semi-trailer.

Though Saturn has 31 moons (at last count), there is only one of a significant size and that is Titan. Titan is the second largest moon in the solar system, at 5,150 km diameter. It also is larger than Mercury, Pluto and our Moon.

So, what will you see of Saturn in binoculars?

Well, as the saying goes, it takes two to tango. So there will be no 'Dance of the Moons', even though you may just be able to spot Titan as a tiny speck of light off to Saturn's side. Titan orbits Saturn every 16 days so you may be able to see it on opposite sides of Saturn every 8 days.

Saturn itself, through binoculars, will look like a yellow star with jug ears, or to put it more scientifically, its disk will appear stretched sideways into an ellipse. Unfortunately, the magnification of binoculars is not enough to see the rings separately from the planet's disk, but you definitely will be able to see that it is not a star, and you will also know that what light you are seeing is more from the rings than the planet itself.

A Few Facts:
- Saturn has a diameter of 120,536 km, approximately one twelfth that of the Sun.
- Its gravity is 1.15 that of Earth.
- Its average distance from the Sun is 1,430 million km, with a minimum of 1,347 million km, and a maximum of 1,507 million km.
- Saturn orbits the Sun every 29.5 years.
- Saturn's "day" is 10 hours 39 minutes, only slightly longer than Jupiter's. That is, it is the second fastest rotating planet in our solar system.
- The moon Titan is the only moon in our solar system to have a significant atmosphere (though not the type you would want to breathe). The nature of this hydro-carbon 'soup' has led many to speculate on the possibility of life on Titan. However, if there is life there, it certainly would not be like anything we could imagine.

Uranus and Neptune:

These two gas planets (the 7th and 8th from the Sun) are actually visible in binoculars if you know exactly where to look.

Uranus has an average distance from the Sun of 2,900 million km and can reach magnitude 5.5 and so it is marginally visible to the naked eye as a very faint blue-green star. But it is best viewed in binoculars as a slightly brighter blue-green star.

Neptune has an average distance to the Sun of 4,500 million km and is magnitude 7.8 at its brightest – just viewable in binoculars, also as a bluish-green star.

To find both these planets (which can be boasted of as a significant achievement to your friends), you will need detailed star maps to locate them within the myriad of stars about them. Again, such charts can be obtained from the ubiquitous Astronomy 200X and the mainline astronomy magazines and their websites.

A Few Facts:

- Uranus has a diameter of 51,118 km
- Uranus orbits the Sun in 84 years
- Uranus has (at last count) 27 moons
- Neptune has a diameter of 49,528 km
- Neptune orbits the Sun in 164.8 years
- Neptune has (at last count) 13 moons

The Stars:

There are stars, and there are stars. Or, to paraphrase that old advertisement, "Stars ain't stars, Sol". One thing they have in common. They are large balls of mostly hydrogen gas - our Sun, at 1.4 million km diameter, is a medium size star – glowing incandescently as a result of the huge release of energy from their central core where a continuous process of hydrogen fusion into helium is taking place. After that, they can be very different.

The science of what makes stars so different from each other is very complex and well beyond the scope of this book. Also, it's not really necessary for our purposes. If you are really interested, there are plenty of good books on the subject (astrophysics, the life cycle of stars) in the libraries.

Suffice to say, stars range from the small (dwarfs) to mid-size (like our Sun) to large (giants) to the massive (super giants). They also range in colour and temperature. (Note: To astronomers, a star's colour and temperature are virtually interchangeable. The colour tells the astronomer the star's temperature.) They go from super-hot (blue) via blue-white, white, white-yellow, yellow, orange to red (cool). This corresponds to temperatures ranging from 40,000°C (blue) to 3,000°C (red). (Our Sun is a main sequence yellow-white star with a surface temperature of 5,500°C.)

To better describe the colours (or temperatures) of the stars, astronomers have devised a numbering system which, due to a multitude of revisions over the years, has become a little scrambled in its order. They allocate a letter to the temperature class of the star. For example, our Sun is a G class star.

The lettering sequence runs, from hottest to coolest, O-B-A-F-G-K-M-N. To help them remember this jumble of letters, they use a mnemonic: "Oh, Be A Fine Girl, Kiss Me Now."

So although, when using binoculars or a telescope, a star still looks just like a star, it is sobering to know what is going on at that point of light. The colour will generally tell you. Is it a cool star, living out its last years before it fades away into a cinder or explodes into a cataclysmic supernova? Is it a hot star, newly born (relatively speaking), about to lead a long stable life in the main sequence, or is it going to burn itself out furiously and fast, also ending in a supernova?

And how big and close is it, how long did it take its light to reach our eyes? Looks generally are deceiving. A bright star may be an average size star that happens to be relatively close, while a fainter star may be a super-brilliant supergiant that is a long way away.

These and many other factors keep a star from being "just a star".

Special hint: There is an easy trick to help you better see the colour of a star in your binoculars. Pin-points of colour are sometimes difficult to identify. So with your binoculars on the star, slowly use the focusing knob to slightly defocus the image. This will cause the star to become a larger blurred image. With the star's colour smeared over the larger surface, it is a lot easier to see if the star is red, orange, blue etc. Having decided its colour, refocus the binoculars to give the sharp image again.

Multiple Stars:

We are very fortunate on Earth that our Sun is a single star. By that I mean the Sun is not gravitationally linked to another star, because if it was, the evolution and sustaining of life on Earth would be highly unlikely, if not impossible.

You might ask, why do I raise this point? The answer is that well over 50% of all stars in our Galaxy (and, we might assume, other galaxies) exist in multiple star systems. That is, two, three or more stars gravitationally bound together, orbiting each other in near-circular or eccentrically elliptical orbits, very close together, or very far apart. Sometimes two stars will orbit each other while a third will orbit around these two. (Alpha Centauri is a close example of this.) In some delightful cases, a pair of orbiting stars will orbit another pair of stars, making a double-double, or a quadruple. And so it goes on.

When two stars orbit each other, they are called a "binary star". This is to distinguish them from the other type of double commonly seen, called an "optical double". An optical double occurs when two stars, totally unrelated gravitationally and probably vast distances from each other, just happen to be aligned from Earth's viewpoint and seem to be very close.

Some optical doubles can be seen with the naked eye, many with binoculars. They have no true scientific significance, but often offer a very attractive sight, especially if the stars are of a different colour or apparent brightness. Later in this book, I shall refer to double stars either as "binaries" or "optical doubles" to distinguish the type of double (or multiple) star. Similarly, if there is a close third star, it may be a genuine orbiting triple (a trinary?) or an optical triple. You can have all the combinations of true binaries and optical extras.

When viewing binary or multiple stars, it is worth keeping in mind the distances between the stars and the sometimes vast periods of time for them to orbit each other. For example, the twin stars of Alpha Centauri are at a reasonably close distance, varying from 10 AUs (i.e Sun-Saturn distance) to 35 AUs (Sun-Pluto distance) and they take 80 years for each orbit. But their third companion, Proxima Centauri (a red dwarf star not visible in binoculars) is 0.1 light years from them (65,000 AUs) and takes about one million years to orbit its binary companions. This is not uncommon for many multiple star systems.

Variable Stars:

A variable star is one whose brightness waxes and wanes on a regular basis and this may be for a number of astronomical reasons, usually associated with the activity of a giant star during a special stage in its life cycle. This is a very large subject within the science of astronomy and is not addressed in this book. (Again, check the libraries.) However, a few of these stars are identified for your observation in the Tour of Constellations as their variation of magnitude is easily identified by the amateur through binoculars and can be fascinating to monitor.

One particularly interesting variation on the binary theme is the "eclipsing binary", which is a special type of "variable star". This is an exciting type of variable star caused when one of the two stars in a binary system passes in front of (or behind) the other star. This leads to a dramatic reduction in the brightness of the star and when measured can allow the calculation of the period of the binary star's orbit and the relative size of the two component stars. This is usually the first indication that the star is a binary, because the companion star is either too close to resolve (that is, split apart visually) or just too dim to see in the light of its brighter companion.

The most famous of all eclipsing binary stars is Algol, in the constellation Perseus, as it was the first variable star to be identified as an eclipsing binary.

The Milky Way:

With the exception of the two Magellanic Clouds seen from the southern hemisphere and possibly the Andromeda Galaxy seen low in the north, everything you see in the sky with the naked eye is contained within our local galaxy, the Milky Way. Every star, every nebula, every star cluster.

Of course, most people think of the Milky Way as that band of white that arches over the sky on a dark night. That is where the Milky Way got its name. But astronomers have extended the name Milky Way to apply to our whole galaxy. As a matter of interest, astronomers also refer to our galaxy, the Milky Way as The Galaxy.

So what exactly is that band of white, the original Milky Way?

Imagine it is night time and you are in the centre of a large unlit football oval which is covered by a layer of fog about 3 metres thick. If you look up towards the sky, you will be looking through about 1 metre of fog. That won't really obstruct your view of the stars in the sky, though you will probably see a few fog particles in between.

However, if you were to look towards the edge of the oval, directly into the fog layer, all you will probably see is ...fog. Lots and lots of fog particles. You won't see the fence nor the trees beyond.

Now imagine that the layer of fog was actually a huge fried egg made up of a few hundred billion stars and that our Sun and Earth are located mid-thickness in the egg-white about two-thirds the way out from the yolk's centre. The egg-white is about 2,000 light years thick and 100,000 light years in diameter. If you looked out of the egg-white through the 1,000 light years of stars, you would be able to see these stars and nebulae in the egg-white on the way and then, with binoculars or a telescope, those galaxies (other 'fried eggs') outside our own.

But if you looked along the egg, you would be looking one way into 20,000 light years of stars, or the other way towards the centre, 80,000 light years of stars. The stars would appear very dense, just like the fog along the ground.

And that's what the arch of the Milky Way is – the thick 'fried-egg' of our Galaxy. When you look towards the constellation of Sagittarius, you are looking towards the 'yolk' of the egg.

On a clear dark night, it is a wondrous thing to scan the Milky Way with binoculars. Almost everywhere you look you will see strings and knots of stars, clusters of stars both loose and tight, dark lane ways between the thick layers of stars. And remember, when you see these expanses of white, it is not smoke or gas. It is an unbelievable number of individual stars going on for light year after light year.

Open Clusters:

One of the major advantages of binoculars over telescopes is the observation of Open Clusters of stars. This is because many of these clusters are such large objects in the sky that the magnification by a telescope brings them too close and you can't see the cluster for the stars. The 7x or 10x magnifications of binoculars, however, is just right and the cluster often just fills the binocular's field of view, presenting a beautiful sight.

Other open clusters are further away, making them smaller and fainter. These can be a challenge to observe in binoculars, sometimes just appearing as a faint speckle of 'star dust'.

What are Open Clusters? Firstly, all the ones we can see in binoculars and amateur telescopes are contained within our Galaxy. Generally, they can consist of as few as 20 and up to hundreds of stars, grouped together in deep space and all travelling together, like a school of fish or a flock of birds. It is believed the cluster's stars all originated from the same star nursery, a huge gas cloud, and when the stars formed millions or billions of years ago, the solar radiation from the stars 'blew away' the remaining gas so that the stars appear clear and bright in space.

The shape of the clusters is perfectly random but sometimes, by chance, they form intriguing patterns which like clouds in the sky, suggest certain objects. Unlike Earth's clouds, however, these shapes remain the same for many thousands of years. Eventually as the cluster's stars drift gradually apart, the perspective will change and so will the pattern. But not in your or my life span.

A large number of Open Clusters are plainly visible to the naked eye, like the Pleiades or Hyades, while most are visible to the naked eye as faint smudges of light, not quite stars but… when you turn your binoculars on them, they snap out at you in their characteristic shape.

After a while you will get to know most of them and you can start a night's observing by jumping around the sky from one cluster to another, ticking off their names or numbers as you go. It's a neat party trick if you have friends over.

Globular Clusters:

Rating very high on the '*Wow!* Factor' scale is the fact that, with your binoculars, you can view a collection of stars that are the oldest stars in the Universe. Though that may seem odd, that is exactly what you are doing when you focus on these fuzzy objects in the sky called 'Globular Clusters', or 'globs' for short.

Our Galaxy has over 150 of these globular clusters orbiting in a spherical halo. It has been shown that most other galaxies have their own population of globular clusters.

Each glob is of a spherical shape, looking like a ball of fairy floss or like a swarm of moths around a street light. The number of stars in each glob varies from as few as 100,000 up to a few million. But the common factor to them all is: they contain the oldest stars in the Universe. Though some of their stars are younger, the majority are old dying stars, created at the earliest days of the formation of the Universe and our Galaxy.

The stars in the globs are thought to be more closely packed than elsewhere in our Galaxy. The average distance between stars in our region of the Galaxy is about 4 light years, whereas in a glob, it is calculated to be about 1 light year. However, at the very heart of the glob, it is believed the distances between stars could be counted in light weeks. Still a lot of space in between, but very close.

Just try to imagine what the sky would be like from a planet around one of these stars. It would be full of thousands of stars even brighter than Sirius, our brightest. It is likely that there would never be a dark night, as the sky full of bright stars would light the night. It is likely that the occupants of such a planet would never even know there is a universe outside their own galaxy – the local globular.

Each glob appears slightly different, either because of the smaller or larger number of stars, or because of a different density of concentration at the centre. Some globs have a very dense core with a gradual thinning out of the stars towards its edge. Some have little concentration in the centre, appearing roughly evenly spread across its width.

As an average, globs are about 150 light years in diameter, though it's not easy to define where exactly in space they end.

Because of their large size, they are visible in binoculars even at great distances. For example, the King of the Globular Clusters, Omega Centauri, is easily visible in binoculars (it can be spotted with the naked eye as a 'star') and it is 17,000 light years away.

Just like the open clusters, the locations of the binocular visible globs can be easily remembered and they form part of the sky tour. Finding each one is like greeting an old friend. Emphasis on 'old'.

Nebulae:

The word 'nebula' comes from a Latin word meaning "cloud". In deep space, these 'clouds' contain gas - mostly hydrogen and helium – and dust, and span vast distances, often hundreds of light years in diameter (with some up to a thousand light years diameter). Some parts of these clouds are more dense than others and gravity causes these denser parts to compact more and more until – Bingo – you have a star, a huge ball of hydrogen gas with a 20 million degree nuclear furnace at its centre.

Nebulae (the plural of nebula) are the birthplace of stars. Some nebulae, photographed by the Hubble Space Telescope, have revealed stars in the actual process of 'lighting up'. Some nebulae, visible in binoculars, contain stars which are a mere tens of millions of years old.

That is a difficult concept to grasp in our minds. We cannot imagine something as large as our Sun starting from a large cloud of gas, compressing into a ball over one million km in diameter, igniting a nuclear fusion reaction at its core and turning into a blazing star. The mind boggles. But that is what is happening in untold nebulae out there.

And when you look through your binoculars at a faint hazy patch of light, like a streetlight seen through a fog, you will be seeing baby stars still in their nursery, possibly waiting for their sibling stars to be born. **Wow!**

Galaxies:

Galaxies are a major challenge for binoculars of 50mm aperture. (The magnification is irrelevant in this case, it's the light gathering aperture that counts.)

The faintest magnitude visible to the human eye (mag. 6) is extended to magnitude 9 with 50mm aperture binoculars. Unfortunately, the vast majority of galaxies have magnitudes of 10+ (that is, fainter than magnitude 9) and can't be spotted in those binoculars. (Of course binoculars with larger apertures, say 70 or 80mm, can get up to mag. 10.)

But... there are some out there brighter than mag. 9 and it is fun to try and find them. If you can catch one, remember what you are seeing. A vast island of stars, one of countless billions of islands in the archipelago of the Universe.

Each galaxy contains, on average, hundreds of billions of stars, all orbiting in a huge catherine wheel structure about 100,000 light years in diameter. Our Milky Way Galaxy is a very ordinary spiral galaxy like most to be seen out there.

Galaxies fall into four basic categories: Spiral, Barred Spiral, Elliptical, and Irregular. Each of these, however, depending on their specific shape, are further sub-categorised. Our Milky Way has always been thought to be a classic Spiral, but recent observations suggest it may actually be a Barred Spiral. The jury is still out on that issue.

Also, spiral galaxies can look completely different depending on their orientation to our line of sight. For example, we may be looking straight down on top of it so it appears as a large circular shape. This is the so-called 'face-on' spiral. Or we can see it from its side, so it looks like a narrow cigar – the so-called 'edge-on' spiral. Or it can be anywhere in between these views. This all adds to the fascination of what you would be seeing.

It is important when trying to view a faint galaxy to ensure your eyes have adjusted to the dark. Give at least 10 or 15 minutes after being in any form of light for your eyes to adapt to total dark.

Even then you may be staring straight at a wispy galaxy and still not be able to see it. There is a technique called "averted vision" that helps you to see that elusive galaxy. That is, don't look directly at it, but focus your eyes off to the top of the field of view and you will see it out of the periphery of your vision. This is not magic. It has to do with the fact that the outer parts of your eye's retina are more sensitive to light.

It's like seeing a thing out of the corner of your eye, but when you look straight at it, it's not there.

Individual galaxies that can be seen with binoculars are listed in the main part of this book. You need only remember three things:
- Adapt eyes to the dark
- Averted vision
- Billions of stars in other Milky Ways.

Constellations – Road Maps in the Sky:

It might surprise you that there are 88 officially recognised constellations in the sky, covering all of the Northern and Southern Hemispheres. Many of these are well known names based on ancient mythology such as Orion, Scorpius, Sagittarius etc.

However, the ancient astronomers were not familiar with the Southern sky, so a large number of constellations seen from the Southern Hemisphere were named by the intrepid navigators in the 1600s and have slightly more mundane names, like Horologium (the Pendulum Clock) and Norma (the Set Square).

At the end of the day, the sky is carved into 88 discrete areas, like the map of Sydney on the inside cover of a City Road Map book, only the sky's areas are anything but equal nor rectangular. Each area represents a constellation.

However, there are more stars in that area than those that make up the main shape of the constellation. To say that a star, cluster, nebula or galaxy appears in constellation 'X' is to say that it can be found in that area of the sky enclosed within that constellation's boundary.

So, we are almost ready to start hunting for "binocular objects", but before we do, some tips on identification jargon. (Where would we be without jargon?)

Star Names:

The vast majority of stars that will be identified will comprise of a Greek letter followed by the abbreviated name of the constellation. e.g. δ Bootis. (See the Table below for the list of Greek letters.)

As δ (delta) is the 4th letter in the Greek alphabet, this would normally (but not always) mean that δ Bootis is the 4th brightest star in the constellation Bootes. This is known as the Beyer numbering system.

The Greek Alphabet

Letter	Name	Letter	Name
α	alpha	ν	nu
β	beta	ξ	xi
γ	gamma	ο	omicron
δ	delta	π	pi
ε	epsilon	ρ	rho
ζ	zeta	σ	sigma
η	eta	τ	tau
θ	theta	υ	upsilon
ι	iota	φ	phi
κ	kappa	χ	chi
λ	lambda	ψ	psi
μ	mu	ω	omega

However, where the number of stars in a constellation exceeds the number of Greek letters, some stars have a Roman letter identification (some upper case, some lower) such as r, R, q, Q etc. Also, some stars have been identified with simple numbers such as 1, 4, 55 etc. This latter system is known as the Flamsteed numbers.

Let's face it... there are a lot of stars up there to be numbered. For our purposes, the numbering system doesn't really matter. What's in a name? It just helps to put a tag on a particular star you want to find.

Constellation Names and Abbreviations:

There is a tradition that the name of a star, when linked with a number or letter and the constellation's name, takes on a genitive form of the constellation's name, and this is often expressed in text with an abbreviation. Examples of this are: the alpha star in Apus would be described as α Apodis or α Aps; the τ star in Cetus will be τ Ceti, not τ Cetus. This can be further abbreviated to τ Cet.

Here is a complete list of all 88 constellations with their genitive name and accepted abbreviation. The list also indicates which constellations will be found in this book (not all constellations are included for reasons explained at the end of this section). Those **not** specifically included are indicated with an *.

Name of Constellation	Genitive form	Abbreviation	
Andromeda	Andromedae	And	
Antlia	Antliae	Ant	
Apus	Apodis	Aps	
Aquarius	Aquarii	Aqr	
Aquila	Aquilae	Aql	
Ara	Arae	Ara	
Aries	Arietis	Ari	*
Auriga	Aurigae	Aur	
Bootes	Bootis	Boo	
Caelum	Caeli	Cae	*
Camelopardalis	Camelopardalis	Cam	*
Cancer	Cancri	Cnc	
Canes Venatici	Canum Venaticorum	CVn	
Canis Major	Canis Majoris	CMa	
Canis Minor	Canis Minoris	CMi	
Capricornus	Capricorni	Cap	
Carina	Carinae	Car	
Cassiopeia	Cassiopeiae	Cas	*
Centaurus	Centauri	Cen	
Cepheus	Cephei	Cep	*
Cetus	Ceti	Cet	
Chamaeleon	Chamaeleontis	Cha	
Circinus	Circini	Cir	*
Columba	Columbae	Col	*
Coma Berenices	Comae Berenices	Com	
Corona Australis	Coronae Australis	CrA	
Corona Borealis	Coronae Borealis	CrB	
Corvus	Corvi	Crv	*
Crater	Crateris	Crt	*
Crux	Crucis	Cru	

Name of Constellation	Genitive form	Abbreviation	
Cygnus	Cygni	Cyg	
Delphinus	Delphini	Del	
Dorado	Doradus	Dor	
Draco	Draconis	Dra	*
Equuleus	Equulei	Equ	
Eridanus	Eridani	Eri	*
Fornax	Fornacis	For	*
Gemini	Geminorum	Gem	
Grus	Gruis	Gru	
Hercules	Herculis	Her	
Horologium	Horologii	Hor	*
Hydra	Hydrae	Hya	
Hydrus	Hydri	Hyi	*
Indus	Indi	Ind	*
Lacerta	Lacertae	Lac	*
Leo	Leonis	Leo	
Leo Minor	Leonis Minoris	LMi	*
Lepus	Leporis	Lep	
Libra	Librae	Lib	
Lupus	Lupi	Lup	
Lynx	Lyncis	Lyn	*
Lyra	Lyrae	Lyr	
Mensa	Mensae	Men	*
Microscopium	Microscopii	Mic	*
Monoceros	Monocerotis	Mon	
Musca	Muscae	Mus	
Norma	Normae	Nor	
Octans	Octanis	Oct	*
Ophiuchus	Ophiuchi	Oph	
Orion	Orionis	Ori	
Pavo	Pavonis	Pav	
Pegasus	Pegasi	Peg	
Perseus	Persei	Per	
Phoenix	Phoenicis	Phe	*
Pictor	Pictoris	Pic	*
Pisces	Piscium	Psc	
Piscis Austrinus	Piscis Austrini	PsA	*
Puppis	Puppis	Pup	
Pyxis	Pyxidis	Pyx	*
Reticulum	Reticuli	Ret	*
Sagitta	Sagittae	Sge	
Sagittarius	Sagittarii	Sgr	

Name of Constellation	Genitive form	Abbreviation	
Scorpius	Scorpii	Sco	
Sculptor	Sculptoris	Scl	
Scutum	Scuti	Sct	
Serpens	Serpentis	Ser	
Sextans	Sextanis	Sex	*
Taurus	Tauri	Tau	
Telescopium	Telescopii	Tel	
Triangulum	Trianguli	Tri	
Triangulum Australe	Trianguli Australis	TrA	
Tucana	Tucanae	Tuc	
Ursa Major	Ursae Majoris	UMa	*
Ursa Minor	Ursae Minoris	UMi	*
Vela	Velorum	Vel	
Virgo	Virginis	Vir	*
Volans	Volantis	Vol	*
Vulpecula	Vulpeculae	Vul	

Numbering Deep Space Objects:

Deep space objects will usually have a number with NGC in front. e.g. NGC6541 is a globular cluster in Corona Australis. The NGC stands for New General Catalog which is an official world wide listing of astronomical objects. However, as all NGC objects tend to fall into these categories, I personally believe it should stand for Nebulae, Galaxies & Clusters.

Some other objects have a prefix IC which stands for Index Catalogues. These are supplements for the NGC list. For simplicity, the star maps provided in this book only show the NGC or IC number without the NGC or IC prefix.

Messier Objects:

Some deep space objects have an additional identity to their NGC or IC number. This is the Messier Number, based on the famous catalogue by the 18th century French astronomer, Charles Messier. Monsieur Messier was a keen comet hunter and after being misled by these 'non-comet' objects, he catalogued them to make sure he didn't waste time on them again. Little did he know that his name would be immortalised for this list of 'non-comets' and not for any comet he discovered (and he discovered many). The objects will have a number with an M in front. e.g. NGC7089 (a globular cluster in Aquarius) is also called M2.

There are 110 Messier objects. These are very popular with amateur astronomers as they are usually (but not always) easy to find, most with binoculars, the remainder with small telescopes. Some amateurs make it a 'quest' to have observed and identified all 110 Messiers. As they are spread over the northern and southern hemispheres, this is not an easy logistical feat and will involve overseas travel, at least for southern observers.

Magnitudes of Stars:

There will be frequent references to the 'magnitude' of a star or object. e.g. Antares is a 1st magnitude star, or Sirius has a magnitude of -1.47. This system can be confusing to the beginner. However, this is the historical system we are stuck with to describe the brightness of the star or object. After some familiarity, it is quite easy and painless.

Basically, this term is synonymous with a star's 'brightness'. That is, to say a star has a magnitude of 1 means it has a brightness of 1 (whatever that means). What confuses most people is that a star with a magnitude of 2 is actually <u>less bright</u> than a star with a magnitude of 1.

Before we go into the actual magnitude numbering system, let's clear up one point. In general conversation, 'magnitude' means the apparent brightness of a star. i.e. as we see it from Earth. This is different from 'absolute magnitude' which tries to define a star's 'true' brightness if it was seen from a standard (but arbitrary) distance of 10 parsecs (32.6 light years). So, if two stars of identical absolute magnitude were different distances away from us, they will have different apparent magnitudes (or just plain magnitudes).

Now to the numbering system, and this is what confuses most. Historically, Hipparcus and Ptolemy divided the naked eye visible stars into six groups. The brightest were called 1st magnitude, the next brightest (but dimmer) were called 2nd magnitude, etc, and the least bright (and barely visible) were called 6th magnitude.

Like Sherlock Holmes, you notice something unusual here? Yes, the less bright stars have a higher number. Or putting it in reverse, the higher the magnitude, the fainter the star. This system is historical and ingrained. We have to live with it.

This numbering system has been developed over the years to include negative numbers (e.g. –1, –4) and even, with the benefit of precise measuring instruments, decimal numbers. (e.g. Sirius is mag. –1.47.)

Is there any method in this madness? Very much so.

Difference in Magnitude	How Much Brighter?
-1	x 2.512
-2	x 6.3
-3	x 15.8
-4	x 39.8
-5	x 100

The magnitude scale is (for the maths buffs) a logarithmic scale, with each step of 1 magnitude representing a ratio in brightness of 2.512. This odd number is the 5th root of 100. Or, 2.512 multiplied by itself 5 times gives 100. So a difference in magnitude of 5 represents a difference in brightness of 100.

Of course, our naked eyes cannot determine magnitudes to that accuracy. The best we can do is no better than good old Hipparcus. But it is very important to understand what the numbering system means. e.g. if a book tells us that a certain deep space object (a galaxy say) is magnitude 11, that tells us it is 5 magnitudes fainter (1/100th the brightness) than the faintest star our naked eye can see (mag. 6).

It also tells us that when Venus is blazing at mag. -4.47 it is 3 magnitudes (2.512^3 = 15.8 times) brighter than Sirius, the brightest star.

Of course, stars not visible to the naked eye have magnitudes higher than 6. And the galaxies at the far reach of Hubble's light grasp are much much fainter still. A mag. 29 galaxy, a test even for the Hubble or Keck telescopes, is 25 mag. fainter than our humble 4th mag. globular cluster ω Centauri. i.e. 100^5 = 10^{10} (ten thousand million) times fainter.

At the other end of the scale, although our Sun has an absolute magnitude of +4.8, being so close, its apparent magnitude is –26.7.

Get the idea?

Finally, since with 50mm binoculars you should be able to see objects as faint as 9th or even 10th magnitude, this means you can see objects from 15 to 40 times fainter than the faintest naked eye star. That's a huge leap out into space.

One Last Point:

As this is a book for Binocular Viewing for the Southern Sky, it does not cover all 88 constellations. Some northern constellations are viewable from the Southern Hemisphere (especially in Australasia) and, where these have binocular viewable objects, they will be included. Otherwise, they are not.

Some constellations viewable in the Southern sky do not contain any objects of interest for binoculars. In these cases, that constellation is not included in this book unless there is some special feature of interest about a particular star or the constellation itself.

Now... out with the binoculars and on with the chase.

Good Seeing.

SECTION 5.

Practical Tips for Observing
(This will help it work)

Great Expectations

It is important before you start delving into your exploration of the night sky that you understand a few basics of astronomical observing. Like learning to drive a car, at first it seems all too hard and you keep on clashing the gears and stalling and think you'll never get it. Then, after lots of practice and patience, Bingo! You suddenly get the hang of it and wonder why you ever thought it was so difficult. If you are an absolute beginner in finding your way around the sky and pointing your binoculars at unknown objects, you will initially experience that frustration. My advice based on personal experience - DON'T GIVE UP!

Like anything worthwhile doing, it is worth perseverance. Trust me. It will get easier and that's when the pain ceases and the pleasure begins.

At the risk of sounding trite, astronomy, and particularly binocular astronomy, can require the application of that uncommon talent of the Three Ps – Patience, Persistence and Practice.

Now, about the objects you will be hunting down with the aid of this book. Like all things in life, some are easier than others. The pleasure in the easy achievements is the almost instant enjoyment of the experience with relatively little effort. The pleasure in the difficult achievements (which may include some frustrating failures along the way) is the immense satisfaction of eventually succeeding and tucking it away in your mental trophy case. But the next time you attempt to find that object again – guess what? It will be a lot easier. Practice makes perfect.

I stress the above point because, as I said, when you work your way through the constellations and follow the book's direction, it will immediately become obvious that some objects are more easily found than others.

Some will literally leap out at your naked eye, shouting "look at me, look at me". The Moon, Venus, Jupiter, the Pleiades, the Hyades, the Milky Way are good examples of this.

Others, though not obvious to the naked eye, once you direct your binoculars to them, will stand out like the proverbial shag on the rock. Once in your field of view, they seem to be hanging there waiting for you to reach out and grab them. The great globular clusters, the larger open clusters, spectacular nebulae like the Orion, Lagoon and Eta Carinae nebulae, and wide double stars are examples of this.

Then there are the fainter or smaller objects that need to be teased out. The success in capturing these can depend on a number of factors. Your eyesight, the aperture or magnification of your binoculars, the steadiness of your binoculars, the clarity of the sky on the night, the darkness of the sky – whether you are in a more light polluted suburban location or out in the country away from city lights. Obviously, the more light pollution there is, the less likely you will be to see the fainter nebulae, or those objects that depend on the contrast of dark against light such as dark nebulae. When trying to observe a binary star or optical double, your chances of success are greatly improved if you can steady your binoculars. It is amazing how two close stars, undetectable in a shaky image, become splittable in a steady image. It is in the hunting of these more elusive prey that you will need to remember the Three Ps.

However, most of the objects described in this section should be viewable from the average suburban backyard, as long as you get out of the glare of that annoying street light or neighbour's backyard flood light.

Dark Adaptation

I imagine many using this book will feel inspired to leap up from their comfortable lounge chair, grab their binoculars and dash out into the dark backyard to find an object. They will look up into an apparently clear sky and immediately say "where are the stars?" Always remember – when coming out of your lit up house into the dark, give your eyes at least 15 minutes to fully adapt to the dark if you want to see the fainter objects. You might begin to see the brighter stars after a few minutes but to be fully adapted takes about 15 minutes. So it's a good idea, if you want to have a long session of viewing, to not keep dashing in and out of the house to check the book. Take the book out with you but DON'T use a naked torch to read it as it will set your dark adaptation back again.

There's a simple solution. If you don't have a red LED light, take your ordinary torch and wrap a few layers of red cellophane around its end, held by an elastic band. You can easily read the book by this light and it leaves your dark adapted eyes unaffected.

Estimating distance angles

I will often say that an object is so many degrees in a certain direction from another object. (eg μ Gruis is 2° NW of δ Gruis.) This is easy to work out when you remember that the diameter of a full moon is half a degree. So, in the example above, I am really saying that one star is four moon diameters away from the other.

Note: 1 degree (1°) = 60 minutes (60')

It will help to have a few rules of thumb to estimate angular distances in the sky. For example, your common biro pen or pencil is about 1° wide when held out at arm's length, especially if your arm is the average 0.5m (20 inches) long.

It varies with individuals, but it is usually the case that the width of your thumb when held out a full arm length is about 2.5°. Four fingers are about 10° wide.

There is a scale printed down one side of the back cover of this book. If it is held out at arms length (0.5m), it will give you accurate angles to use to follow my directions. It would be worth while using this scale to estimate the angle between the tip of your little finger and thumb when your hand is stretched wide. This way you can easily measure out larger distances in the sky. For example, my finger-thumb width is 25°.

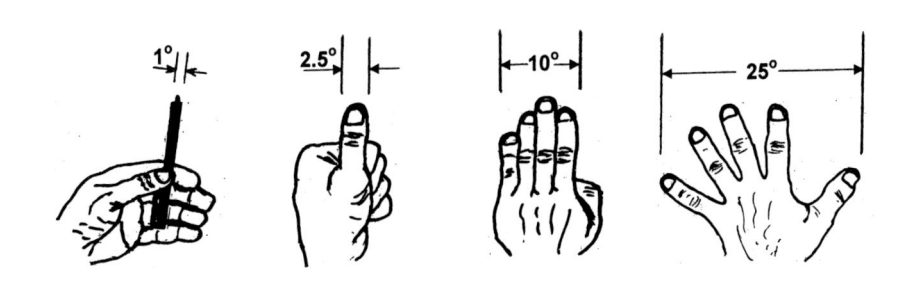

Tips for Steadying Your Binoculars

As is have mentioned earlier in the book, binoculars can become heavy to hold after a while, and the higher magnification they have, the smallest tremor in your hands translates to a significant wobble in the image. This is particularly annoying when you are trying to see small points of light close together, such as Jupiter's moons or binary or double stars. Your chance of being able to see these objects is greatly increased if you can keep the binoculars steady in your hands, or better still, in some kind of hands-free mount.

In this section I describe a few simple techniques to steady your binoculars. Some are pretty basic, probably even intuitive. Others are a bit more sophisticated but would require some effort to obtain or construct. Some suggested mechanical devices may be beyond most peoples ambitions or handy-man skills. However, some of you DIY people may consider them a snap. You decide what works for you and how ambitious you want to be. Personally, I use the most basic methods and they work for me.

Method 1: If you have a house or shed wall nearby and if your field of view will allow it, simply stand next to a corner of the wall and support one side of your binoculars against it. Though still holding the binoculars in both hands, any tremor in them is damped (even totally eliminated) by the steadiness of the wall. This is usually very effective. Unfortunately, this doesn't work well for objects that are high overhead as eaves and such will probably block your view.

Method 2: This is a simple variation of Method 1 above, whereas instead of a wall corner, use the top of a fence post or panel. Again, this eliminates most, if not all wobble. Be aware you may have to explain to your neighbour what you are doing.

Methods 1 and 2, while effective in steadying your binoculars, have limitations as the locations of your target object and available walls and fences may not physically allow comfortable viewing. It is much better if you can use a portable steadying device.

Method 3: This is a very simple device which you can easily make yourself. There is no unique design, just a concept, and you construct it from whatever materials are on hand. Basically, the idea is a long stiff rod with a tee bracket at the top to which you fix the binoculars, and another tee bracket at the bottom which stands on the ground to steady it (the lower bracket is not essential).

Possible materials for the rod are old broom, mop or rake handles, thick dowel rod or the like. You can be very clever and use a telescopic rod, such as an extendable paint roller handle or similar. This allows some adjustment of the length where circumstance require it such as standing or sitting.

An ideal 'ready to use' device is an old mop handle, with the mop removed to leave the flat wooden piece at the end. Binoculars can easily be fixed using elastic straps, velcro strips or a clamp of your own design. Even just resting it on the wooden support works. Keep it simple so the binoculars can be easily mounted or removed as required.

For objects close to the horizon, one could stand with the rod vertical, planted firmly on the ground and you holding on to the binoculars which are steadied by the whole thing. If objects are higher up in the sky, you can sit in a chair and place the base of the rod out in front of you, possibly held to the ground by your feet, while the binoculars on the top cross-piece are pointed upwards at a sharper angle.

The two examples on the right are of the author with a simple extension handle (for a paint roller) and a flat block of wood inserted in the end. In these cases, the binoculars are just resting on top of the block, steadied by the author. This contraption took 20 minutes to make from bits and pieces.

This device can be as crude or as complex as you want it, depending on your wood working skills. One way to make it more comfortable is to have the top mounting piece on a pivot with the rod so that it can adjust the angle of the binoculars more easily. One simple way to do this is use a wing nut arrangement to easily loosen, adjust then retighten the mount. In fact, it is better if not tightened completely as a bit of up-down movement allows for easier searching of the sky and it will still remove any tremor.

Method 4: This is rather sophisticated and recommended only for those really keen who have good handy-man skills. (Or you can buy one ready made.) It is generally described as a parallelogram mount. An example of this device is shown in the picture below, courtesy of a friend in the Macarthur Astronomical Society.

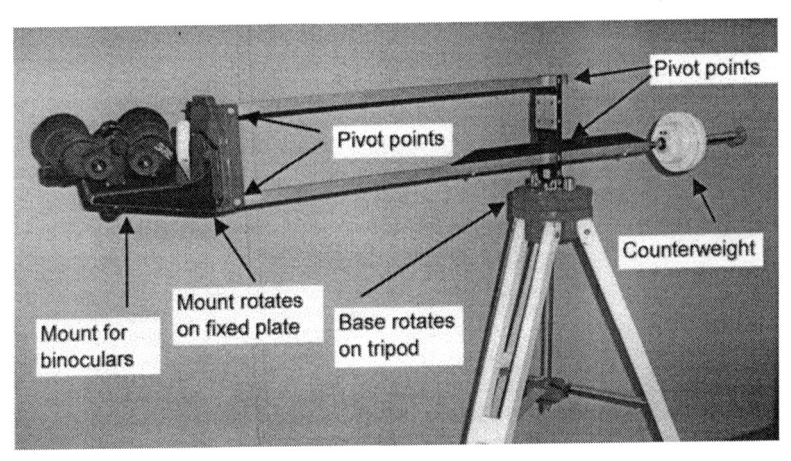

This is shown only as an example of how it can be made. You can design your own using whatever materials you have available.

In essence, you have a robust tripod and on this is mounted a vertical shaft. The shaft is attached to the tripod in a manner that allows it to rotate. To the shaft is attached an arrangement of arms that form a long narrow parallelogram.

The lower long arm is attached to the vertical shaft by a pivot connection and at the end holds an adjustable counterweight. The upper long arm is also attached to the vertical shaft by a pivot connection. The short arm at the other end of these two long arms are also connected by pivots. This means that the parallelogram can be moved up and down easily and though its shape will change, the short arm will remain vertical.

Connected to the short arm is the arrangement to hold the binoculars usually employing a threaded mounting hole provided in the front centre of most binoculars. This arrangement should also allow it to be able to be twisted at a vertical angle to allow the binoculars to tip up or down to suit the direction of the target object.

The advantage of this mount is that once the binoculars are raised to a comfortable height and pointed at the target object (which can be on the horizon, directly overhead or anywhere in between), it will stay there if you let it go, by courtesy of the counterweight. Better still, if you invite someone else to look and they are of a different height, you simply raise or lower the binoculars which moves the parallelogram but it stays exactly on target.

Some detailed views of the construction of the binocular mount are shown in the pictures below. Again, design it to suit your binoculars dimensions and materials available. Note that my friend added a 'rifle sight' to allow him to align the binoculars onto the viewing object.

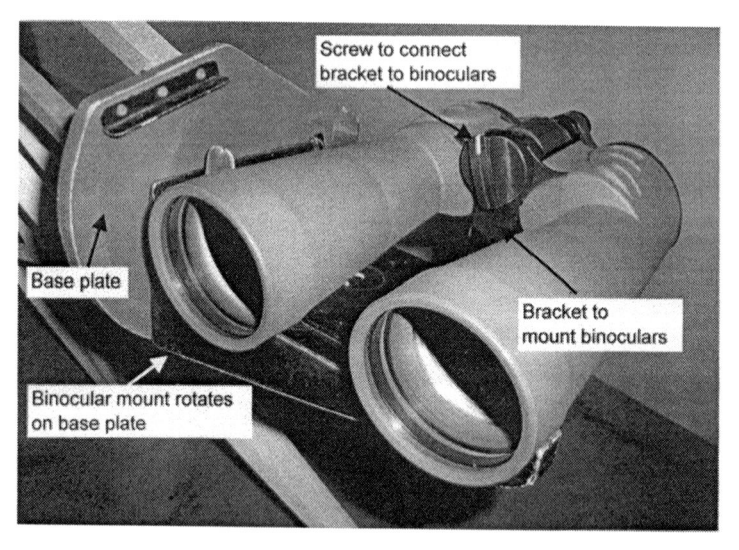

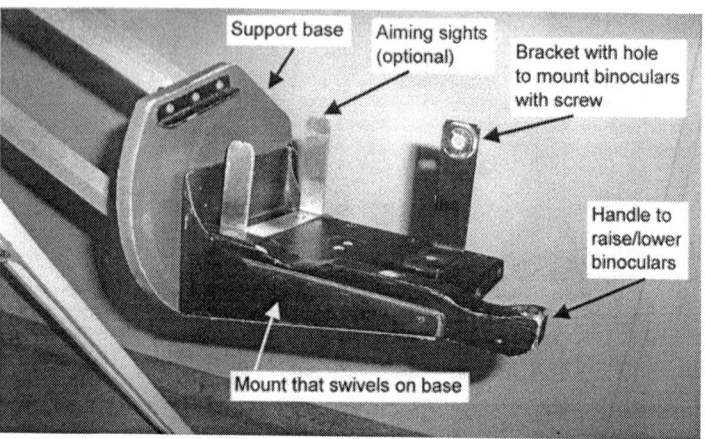

Method 5: The Binocular Chair. There are any number of designs out there for chairs, from the humble Director's Chair to arm chairs, which have a host of ingenious arrangements to allow binoculars to be supported from above and/or behind so that you simply pull them down to meet the height of your eyes and adjust the angle for viewing. This is binocular viewing at its most decadently comfortable. (Of course, you can always use the parallelogram support from method 3 while seated.)

If you want to enjoy this experience, I will leave it to you to research the internet to choose a design. Try searching on "Binocular chair", 'steady binoculars" or "binocular mount". They all give hits of examples of people's designs. Of course you can buy them, but they'll cost a lot more than your binoculars.

I would suggest you start with the humble mop handle or similar and move up from there as the mood (and your creativity) suits you.

Finding the South Celestial Pole

A number of the constellations covered in this book are a bit obscure and directions to find them, or objects within them, are often referenced to the South Celestial Pole (SCP). The SCP is the point in the sky about which the Southern hemisphere stars rotate, the projection of the Earth's axis through the Earth's South Pole. But how do you find that point in the sky that is the SCP?

Unlike the Northern hemisphere, the Southern hemisphere doesn't have a Pole Star (like the Northern's Polaris). That part of the Southern sky is mostly devoid of stars, at least to the naked eye. But there are a number of simple tricks to locate the SCP closely enough for our astronomical observing purposes.

Also, this is a handy trick to know if ever you are out at night and need to find which way is South (or North, East or West). Once you find the SCP, simply drop down straight to the horizon and that direction is South. Obviously, directly behind you is North, to your left is East and to your right is West. This is one of the most basic examples of 'navigating by the stars'.

There are a number of easily identified objects (and they get easier with familiarity) to the South which, when combined into certain pairs and some simple geometry, give you the SCP. The geometry usually involves drawing some imaginary lines or triangle in the sky with your left and right hands (much to the amusement of your family or friends). The method to use will depend on what objects are visible and above the horizon on that night and their orientation in the sky. There will always be least one of the five methods given below available at any time on any night, clouds permitting. Keep in mind, there is a reverse benefit in learning these techniques. Once you know the location of the SCP and you can see one of the other objects, you can find the other object involved in the technique. This is very helpful in light polluted skies.

Method 1: The Southern Cross

If you simply extend the long axis of the Southern Cross (Crux) from γ to α by four times the cross's length, the line will end very close to the SCP.

Method 2: Southern Cross and the Pointers

Locate Crux and the two Pointer stars, α and β Centauri. Use one hand to draw a line through the long axis of Crux, as for Method 1, but don't measure lengths. Use the other hand to bisect at right angles the line between the two Pointers and extend the bisecting line in the same direction. Where the two imaginary lines meet is very close to the SCP.

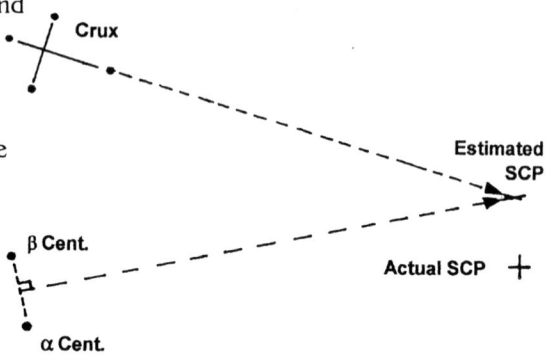

Method 3: Southern Cross and False Cross

If the Pointers are below the horizon or behind trees but Crux is visible, you should also be able to see the False Cross in Carina/Vela. Locate the centres of both the crosses and using your hands, create an imaginary equilateral triangle with the line between the cross' centres as one side and the opposite apex of this triangle on the side nearest the feet of the crosses. This apex is then very close to the SCP.

There are times when the crosses and Pointers are so low they are hidden by the horizon, trees or clouds. At such times, the next two methods are very useful.

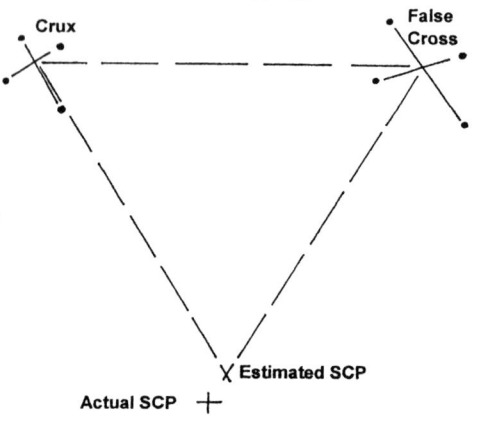

Method 4: Stars Canopus and Achernar.

Looking south, find Canopus, the second brightest star in the sky, in Carina. Then find Achernar, which is a bright white magnitude +0.5 star in Eridanus. (If you haven't already learnt to identify these stars by sight, use the monthly star map which will have them clearly identified.)

Use the equilateral triangle method, with Canopus and Achernar as two of the corners, to find the SCP at the third corner.

Method 5: Large and Small Magellanic Clouds

For similar nights when you could use Method 4, this method is also available. You might be more comfortable spotting the Large and Small Magellanic Clouds (LMC & SMC) than the stars Canopus and Achernar. Once you have located the LMC and SMC, use the equilateral triangle method and find the SCP at the third corner.

Remember, for Methods 3, 4 and 5, always use the orientation shown otherwise your triangle will point in the wrong direction and your 'SCP' will be way off target.

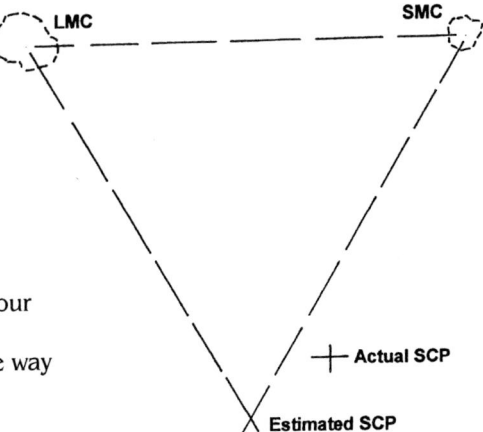

All five methods will be valid regardless of the orientation of the objects in the sky. The orientation shown in the diagrams are typical samples only. The geometry is always the same and will give the same result, as the SCP is fixed in the sky. All that may change is the visibility of the objects involved. A sketch showing the relative positions of all the methods' objects is shown on the next page. Again, depending on the date and time, this whole diagram could be rotated but it will still be relevant.

As a reality check of your spotting of the SCP's location, the angle of the SCP above your horizon should be equal to your local latitude. For example, if you are in the same general latitude as Sydney, you would expect the SCP to be about 35° above the horizon.

With a little practice, the process using any of the above methods becomes second nature, but it is very impressive to friends at outdoor parties.

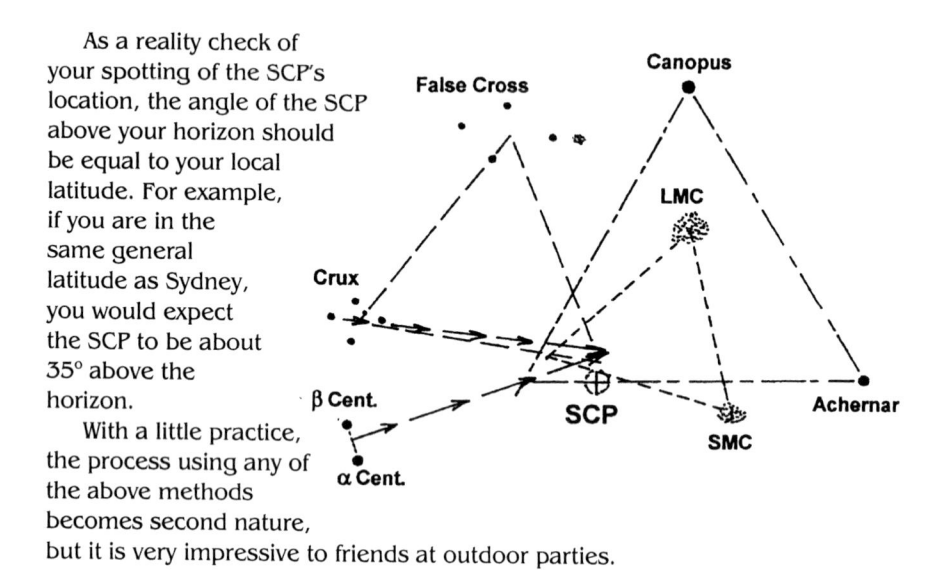

Averted Vision

One last tip. There will be times you will be hunting for a very faint object such as a nebula or a galaxy. You are certain you have the right location and it should be there, right in the centre of your field of view. You can see a suggestion of a smudge of light but it's not that obvious.

There is a simple way to see the object a bit more clearly. It's called 'averted vision'. This is where you don't look directly at the target object but a bit off centre. The reason for this has to do with how the eye works. Briefly, it is made up of around 5 million cones in the centre (sensitive to high brightness objects) and about 100 million rods around the outside (sensitive to low brightness objects.)

If you are looking directly at an object, the outer rods are not coming fully into play and the cones don't show the low brightness object that well. If you can bring the outer rods into play, you will see the object much more clearly.

Traditionally, with a telescope eyepiece, this would be done by simply averting your gaze towards the edge of the field of view. In your averted vision, you would see the target more brightly, out of the corner of your eye as it were. But it is best not to do this with binoculars, as there is a blind spot to one side of each eye, and it would cancel any benefit of the averted vision.

The trick with binoculars is this: avert your gaze upwards. This way both eyes are using the rods at the top of your retinas and there is no blind-spot affect.

So when that faint object is in the centre of your field of view, leave the binocular as it is but raise your gaze towards the top of the field of view. The object should look a lot brighter in your peripheral view. Try it – it works.

SECTION 6.

A Tour of the Constellations
(What and Where?)

In this section, the suggested months for viewing are based on 9pm Eastern Standard Time. Viewing is also possible in earlier months if you wish to stay up later towards midnight or later months if you can get an early start in the winter nights.

My suggested method is:

Step 1 - you first establish what constellations are visible on the night in question using the 'Visibility of Constellations' table on pages 55-57, then;

Step 2 - find the general position in the sky for the chosen constellation by using the monthly maps provided in this section on pages 58 to 81. After that;

Step 3 - refer to the page for the specific constellation you want to find objects in. They are found alphabetically in pages 82 – 156.

This 3 Step process should make use of the book and finding the constellation and objects much easier.

Visibility of Constellations

The following table gives a quick guide to those months when a particular constellation is visible over the horizon and also the months when it is best viewable. These are based on viewing at 9pm Eastern Standard Time.

Visible above horizon

Best viewable

To find out what you can see in a particular month, check the shaded constellations under that month in the table.

Constellation	Jan	Feb	Mar	Apr	May	Jun	Jul	Aug	Sep	Oct	Nov	Dec
Andromeda												
Antlia												
Apus												
Aquarius												
Aquila												
Ara												
Bootes												
Cancer												

Constellation	Jan	Feb	Mar	Apr	May	Jun	Jul	Aug	Sep	Oct	Nov	Dec
Canes Venatici												
Canis Major												
Capricornus												
Carina												
Centaurus												
Cetus												
Chamaeleon												
Coma Berenices												
Corona Australis												
Corona Borealis												
Crux												
Cygnus												
Delphinus												
Dorado												
Equuleus												
Gemini												
Grus												
Hercules												
Hydra												
Leo												
Lepus												
Libra												
Lupus												
Lyra												
Monoceros												
Musca												
Norma												
Ophiuchus												
Orion												
Pavo												
Pegasus												
Perseus												
Pisces												
Puppis												
Sagitta												
Sagittarius												
Scorpius												
Sculptor												
Scutum												
Serpens-East												
Serpens-West												
Taurus												
Telescopium												
Triangulum												

Constellation	Jan	Feb	Mar	Apr	May	Jun	Jul	Aug	Sep	Oct	Nov	Dec
Triang.-Aust.												
Tucana												
Vela												
Vulpecula												

Using the Monthly Star Maps

In this section which covers the Step 2 I mentioned at the start of this Section 6, there is a set of double pages showing the sky's view for each month of the year. On the left page is the view north and on the right is the view south. These drawings are shown for a latitude around 35° south.

Whatever month you wish to view the sky, select that page. To locate a constellation to the north, turn the open page anti-clockwise to bring the northern horizon to the bottom. You will quickly see which constellations are directly north, or to the east or west, and how high they appear above the horizon. The top of the circle is about 70° above the horizon.

To locate a constellation while facing south, turn the book clockwise to bring the southern horizon to the bottom. Again, you will see which constellations are directly south, east or west. When facing south, note that there is a small cross at lower center to indicate the location of the South Celestial Pole (SCP). The top of the circle on the southern view page actually extends beyond 'straight overhead' by another 30°, overlapping the top of the northern view, so some northern constellations can be seen in both the northern and southern views. Choose whichever viewing direction you find most comfortable.

For each month, the constellations' locations are correct for 10pm at the start of the month, 9pm at mid-month and 8pm at the month's end. For Daylight Saving months, you will need to subtract 1 hour from those times.

The view for each adjacent month is 2 hours ahead or behind. For example, if you want to observe at 11pm in July, you can use the chart for 9pm in August. That is, to go ahead 2 hours, then move 1 month forward. If you want to observe at 7pm in July, then go back 1 month and use the chart for 9pm in June. Easy!

So, for the individual constellations explained later in this section (which indicate if they have a southerly or northerly aspect), check what month they are visible, find the right monthly map page, then face the suggested direction. Locate the constellation on the monthly map and then search that area of the sky for the distinctive shape of the constellation shown on the map.

Some of the constellations are small and obscure and may not be easy to find. In that case, find the other distinctive constellations shown on the sky map of that constellation provided in this book and work from there. Once you know from the monthly map roughly where the constellation is in the sky, search that area of the sky until you find it. Some times it will be easy, other times not so easy. That's half the fun. The stars are there, you just have to use the maps in the book to identify them.

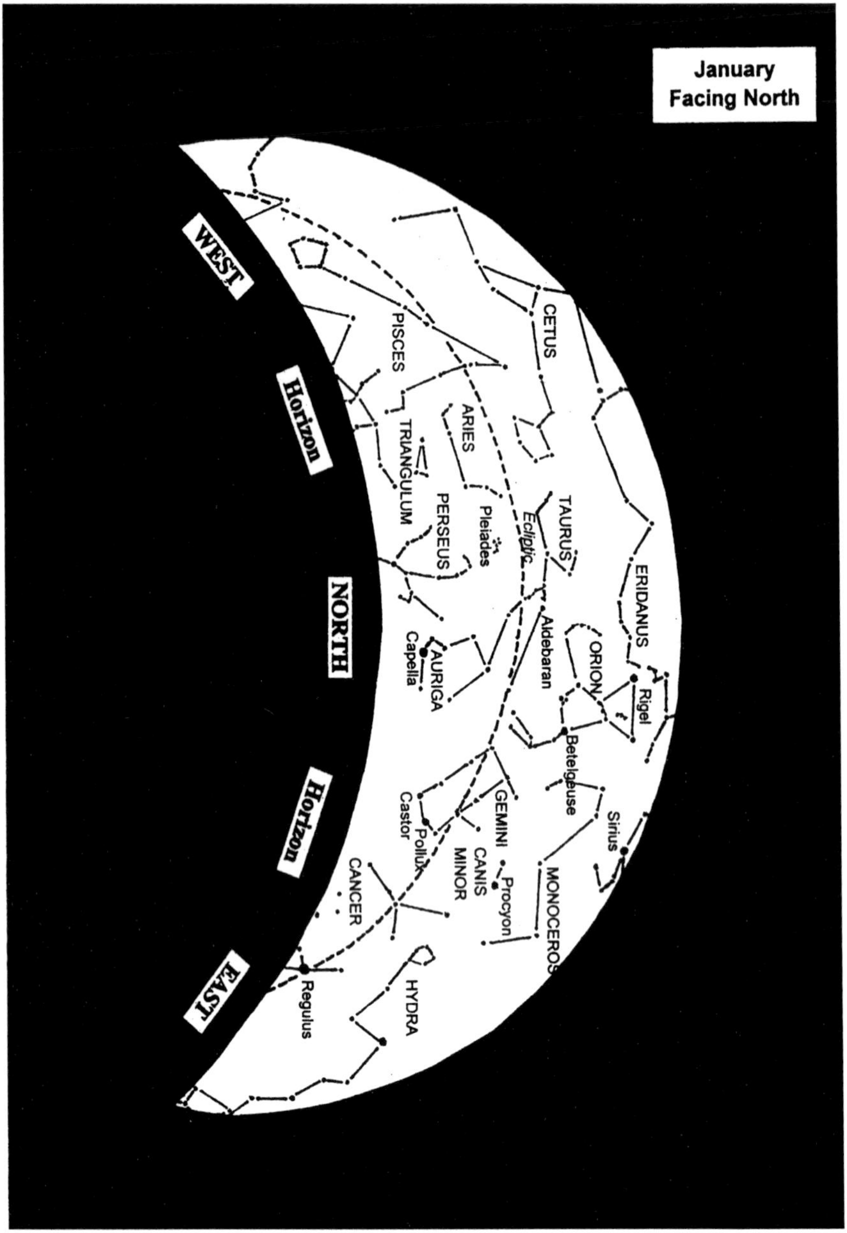
January
Facing North
WEST
Horizon
NORTH
Horizon
EAST
PISCES
CETUS
ARIES
TRIANGULUM
PERSEUS
Pleiades
Ecliptic
TAURUS
ERIDANUS
Aldebaran
Capella
AURIGA
ORION
Rigel
Betelgeuse
GEMINI
Castor
Pollux
CANIS
MINOR
Procyon
MONOCEROS
Sirius
CANCER
HYDRA
Regulus

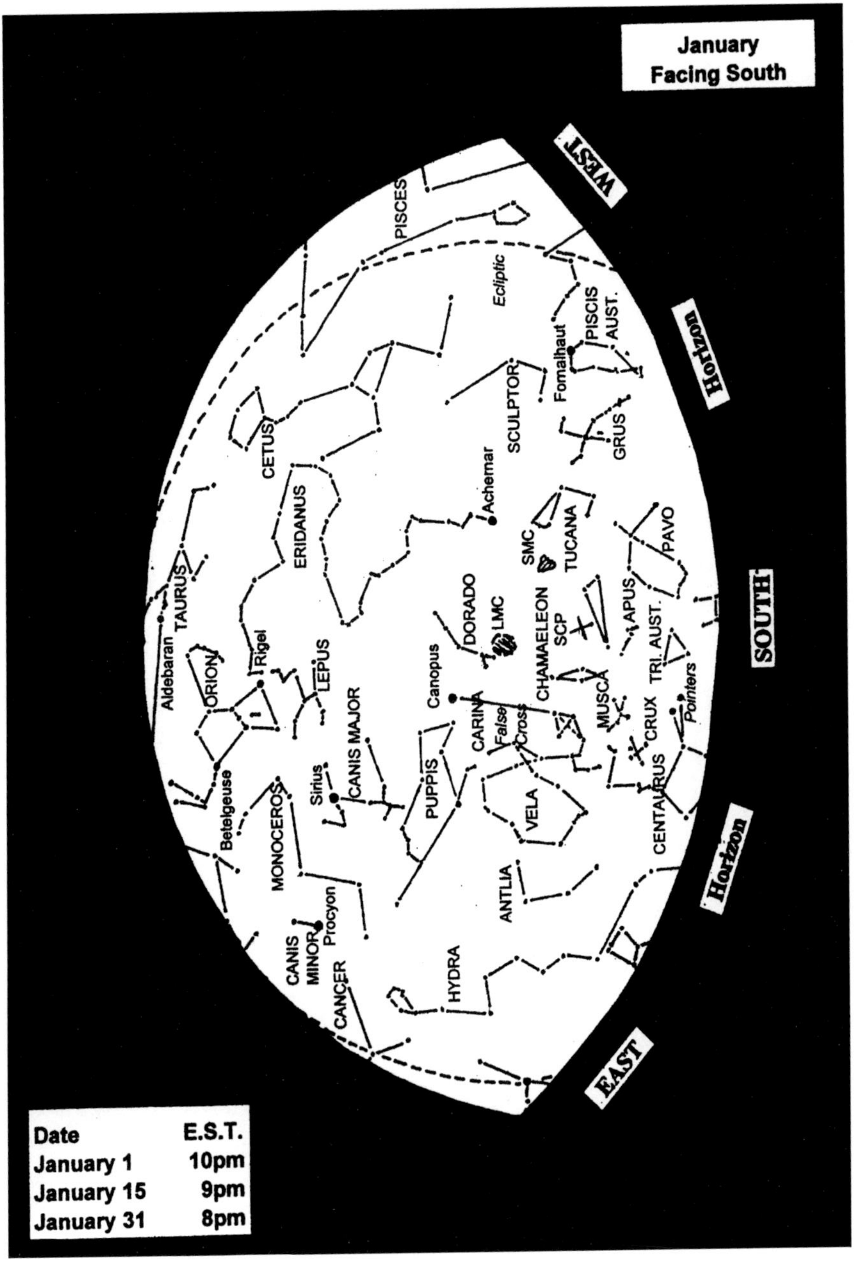
January
Facing South
WEST
Horizon
SOUTH
Horizon
EAST
Date E.S.T.
January 1 10pm
January 15 9pm
January 31 8pm
PISCES
Ecliptic
PISCIS AUST.
Fomalhaut
SCULPTOR
GRUS
CETUS
Achernar
ERIDANUS
TAURUS
Aldebaran
SMC
TUCANA
PAVO
DORADO
LMC
CHAMAELEON
SCP
APUS
TRI. AUST.
ORION
Rigel
LEPUS
Canopus
CANIS MAJOR
CARINA
False Cross
MUSCA
CRUX
Pointers
Betelgeuse
MONOCEROS
Sirius
PUPPIS
VELA
CENTAURUS
ANTLIA
Procyon
CANIS MINOR
CANCER
HYDRA

60

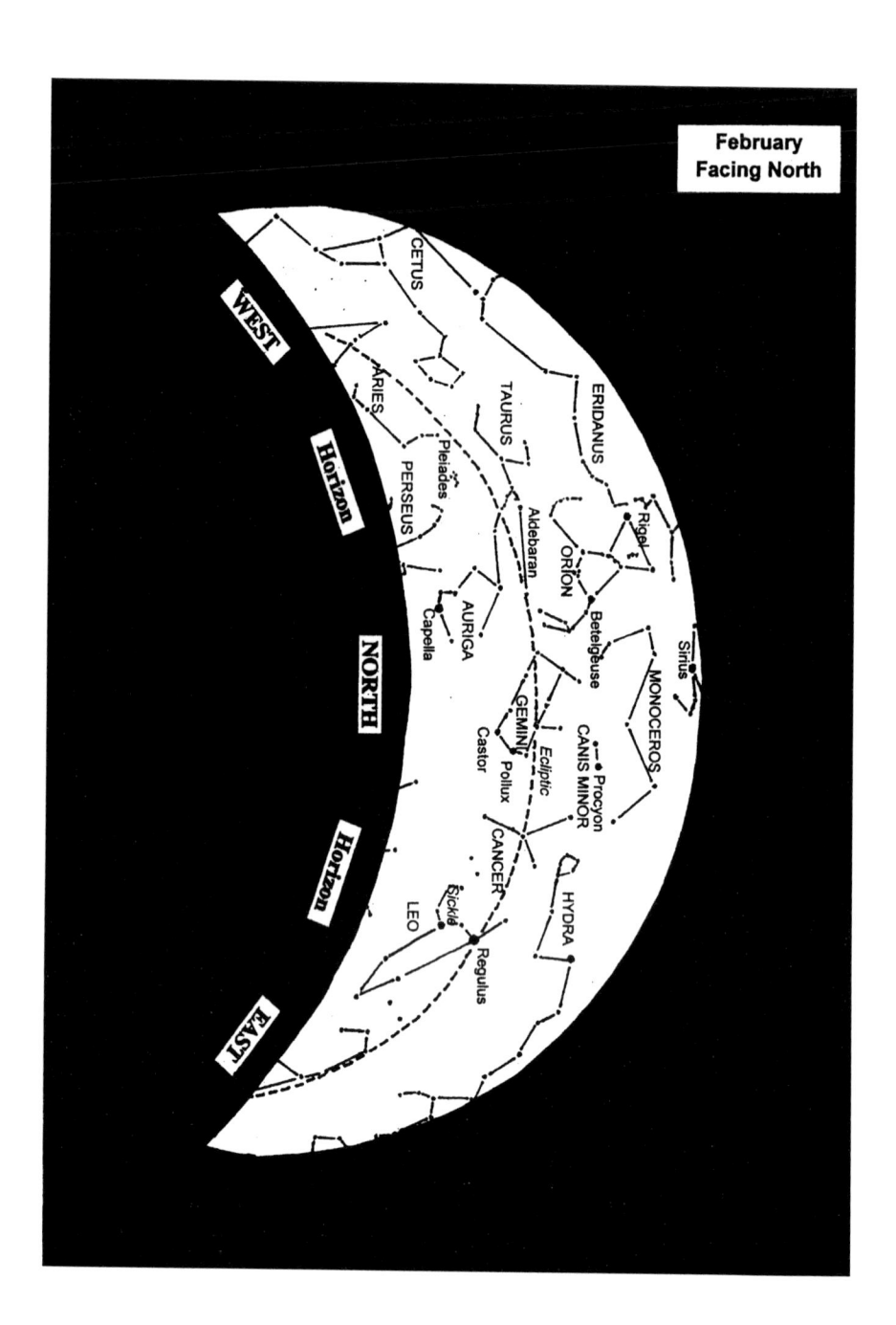

February Facing South

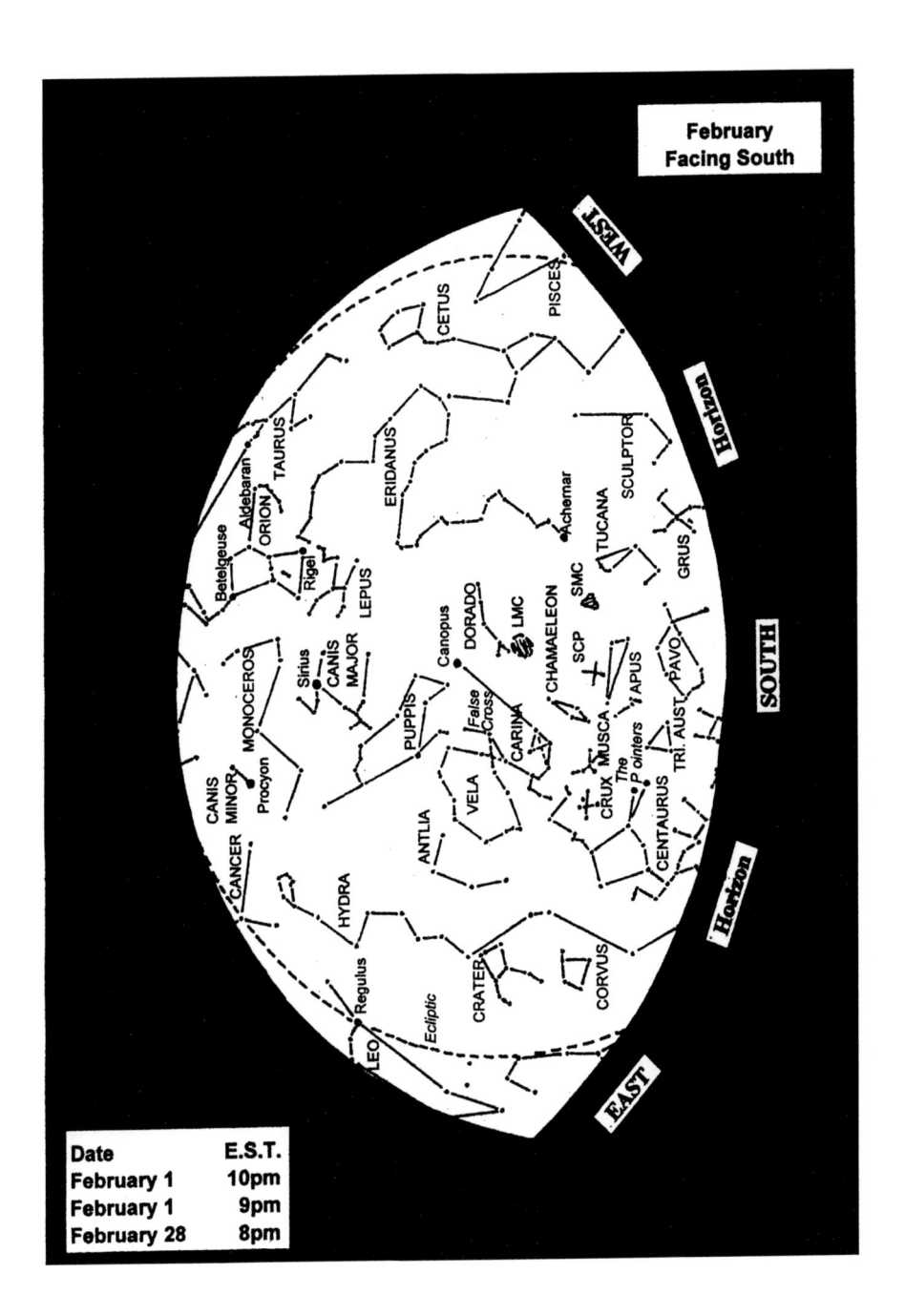

Date	E.S.T.
February 1	10pm
February 1	9pm
February 28	8pm

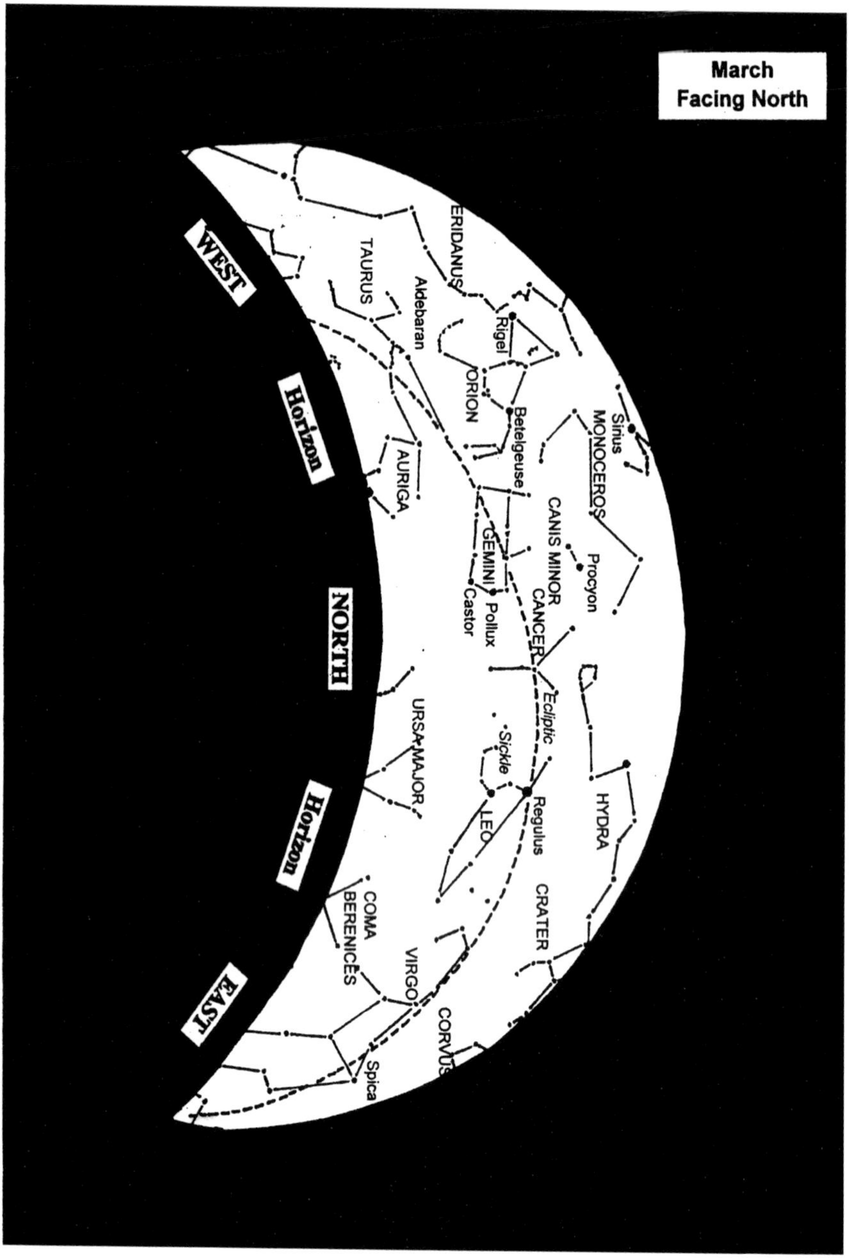
March
Facing North
WEST
Horizon
NORTH
Horizon
EAST
ERIDANUS
TAURUS
Aldebaran
Rigel
ORION
Betelgeuse
Sirius
MONOCEROS
AURIGA
CANIS MINOR
Procyon
GEMINI
Castor
Pollux
CANCER
Ecliptic
Sickle
URSA MAJOR
LEO
Regulus
HYDRA
CRATER
COMA
BERENICES
VIRGO
CORVUS
Spica

March
Facing South

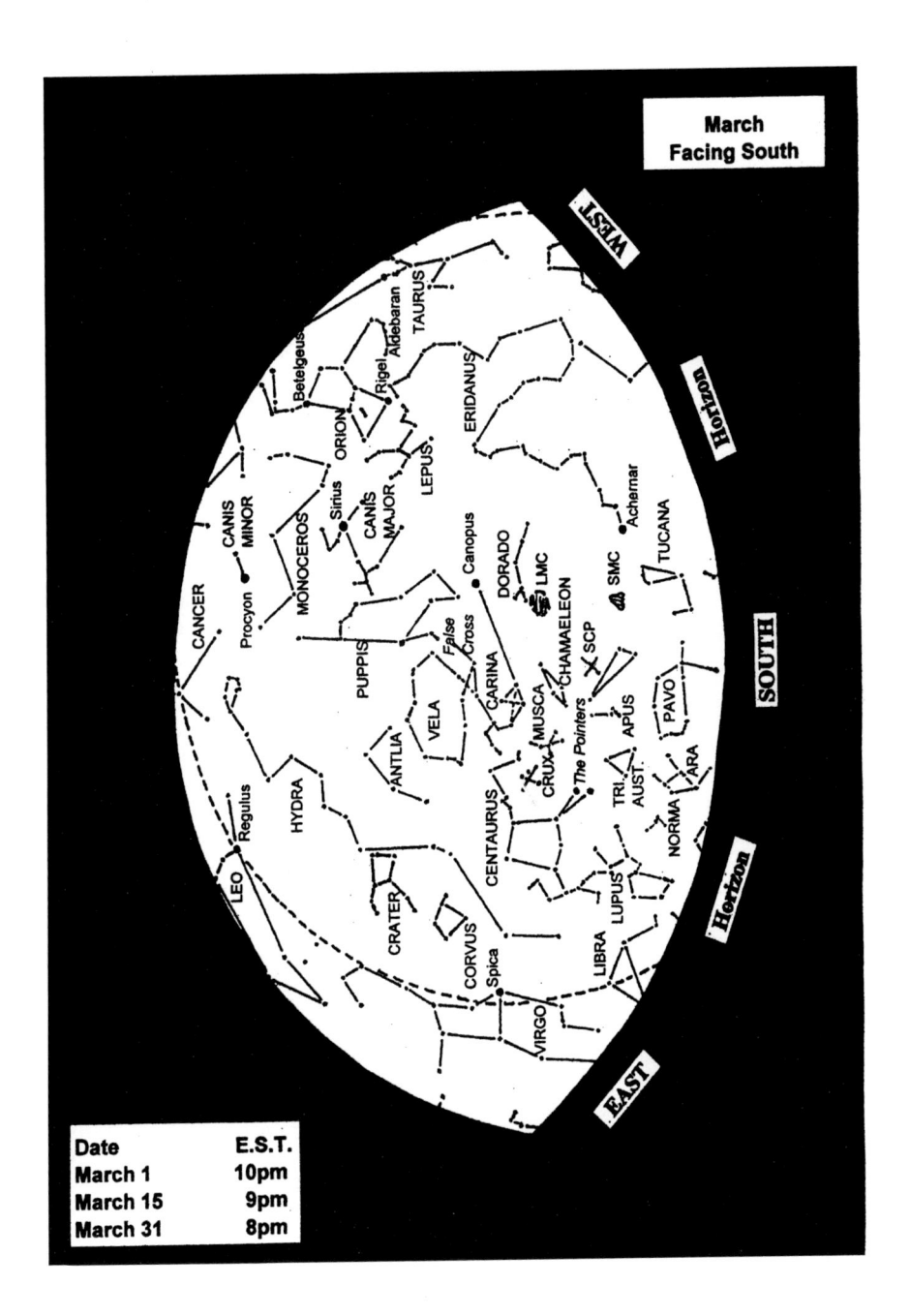

Date	E.S.T.
March 1	10pm
March 15	9pm
March 31	8pm

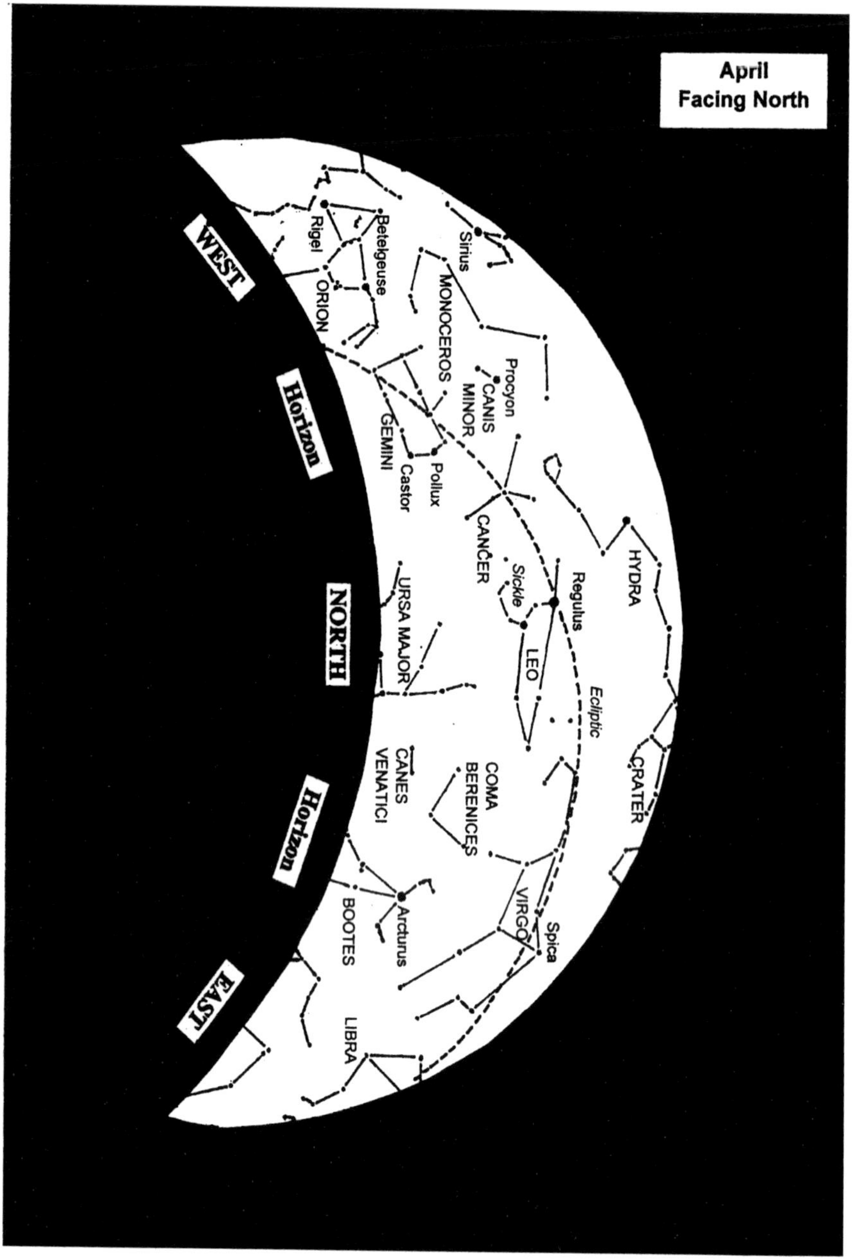
April
Facing North
WEST
Horizon
NORTH
Horizon
EAST
Rigel
Betelgeuse
ORION
Sirius
MONOCEROS
Procyon
CANIS
MINOR
GEMINI
Castor
Pollux
CANCER
Sickle
Regulus
HYDRA
LEO
URSA MAJOR
Ecliptic
CANES
VENATICI
COMA
BERENICES
CRATER
VIRGO
Spica
Arcturus
BOOTES
LIBRA

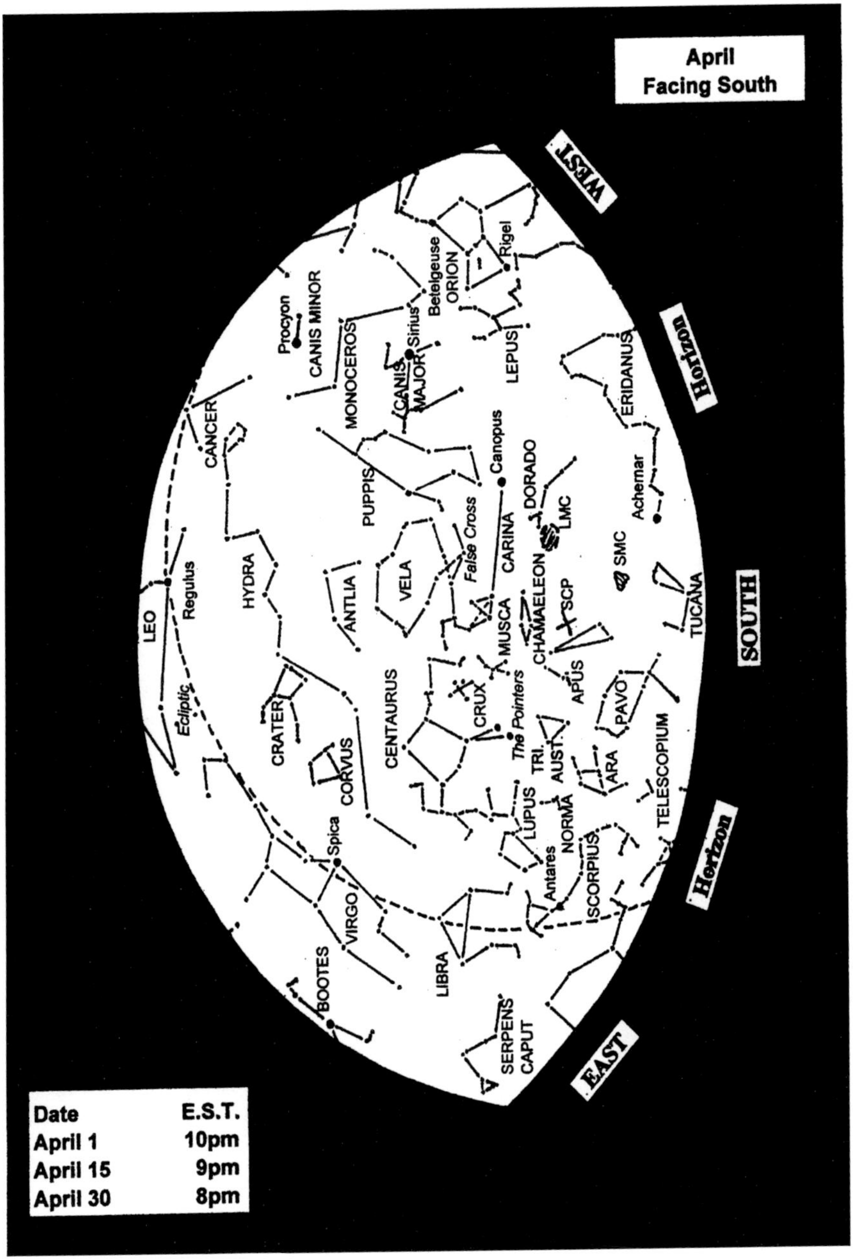
April
Facing South
WEST
Horizon
SOUTH
Horizon
EAST
Date E.S.T.
April 1 10pm
April 15 9pm
April 30 8pm
ORION
Betelgeuse
Rigel
Procyon
CANIS MINOR
MONOCEROS
Sirius
CANIS MAJOR
LEPUS
ERIDANUS
CANCER
PUPPIS
Canopus
DORADO
Achernar
LMC
SMC
False Cross
CARINA
VELA
ANTLIA
HYDRA
LEO
Regulus
Ecliptic
MUSCA
CHAMAELEON
SCP
TUCANA
CRUX
The Pointers
CENTAURUS
CRATER
CORVUS
APUS
PAVO
TRI. AUST.
ARA
TELESCOPIUM
LUPUS
NORMA
Antares
SCORPIUS
Spica
VIRGO
LIBRA
BOOTES
SERPENS CAPUT

May Facing North

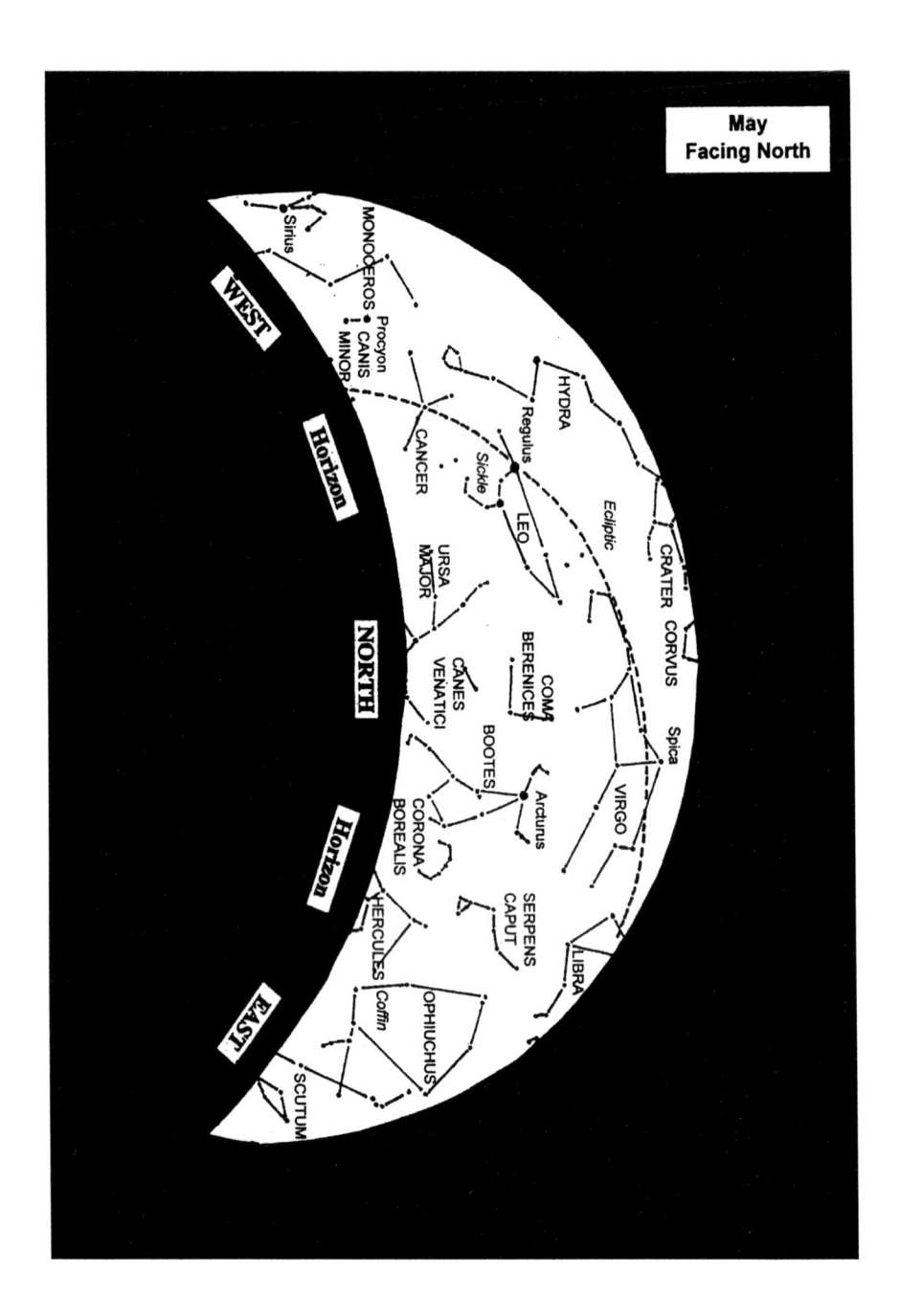

May Facing South

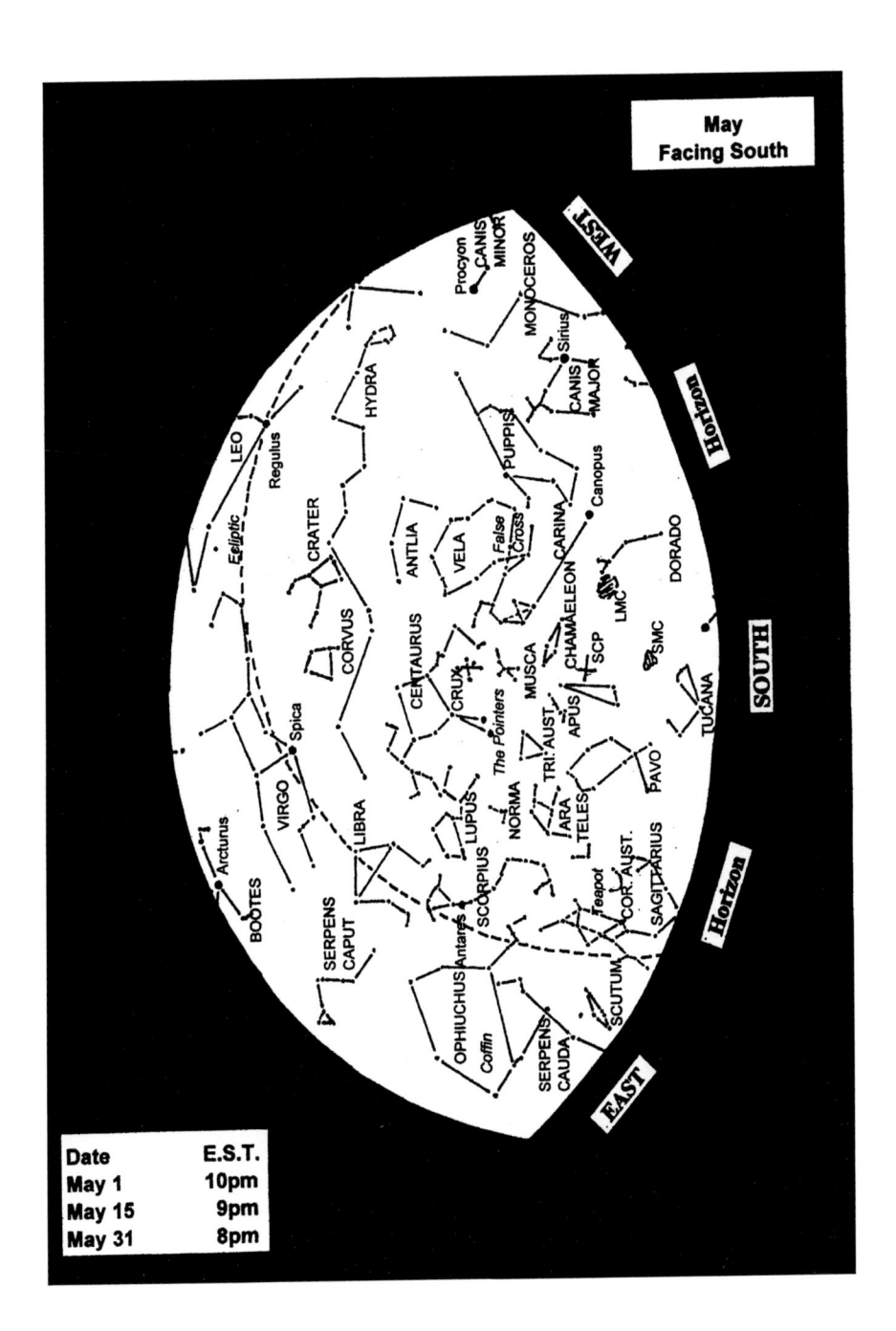

Date	E.S.T.
May 1	10pm
May 15	9pm
May 31	8pm

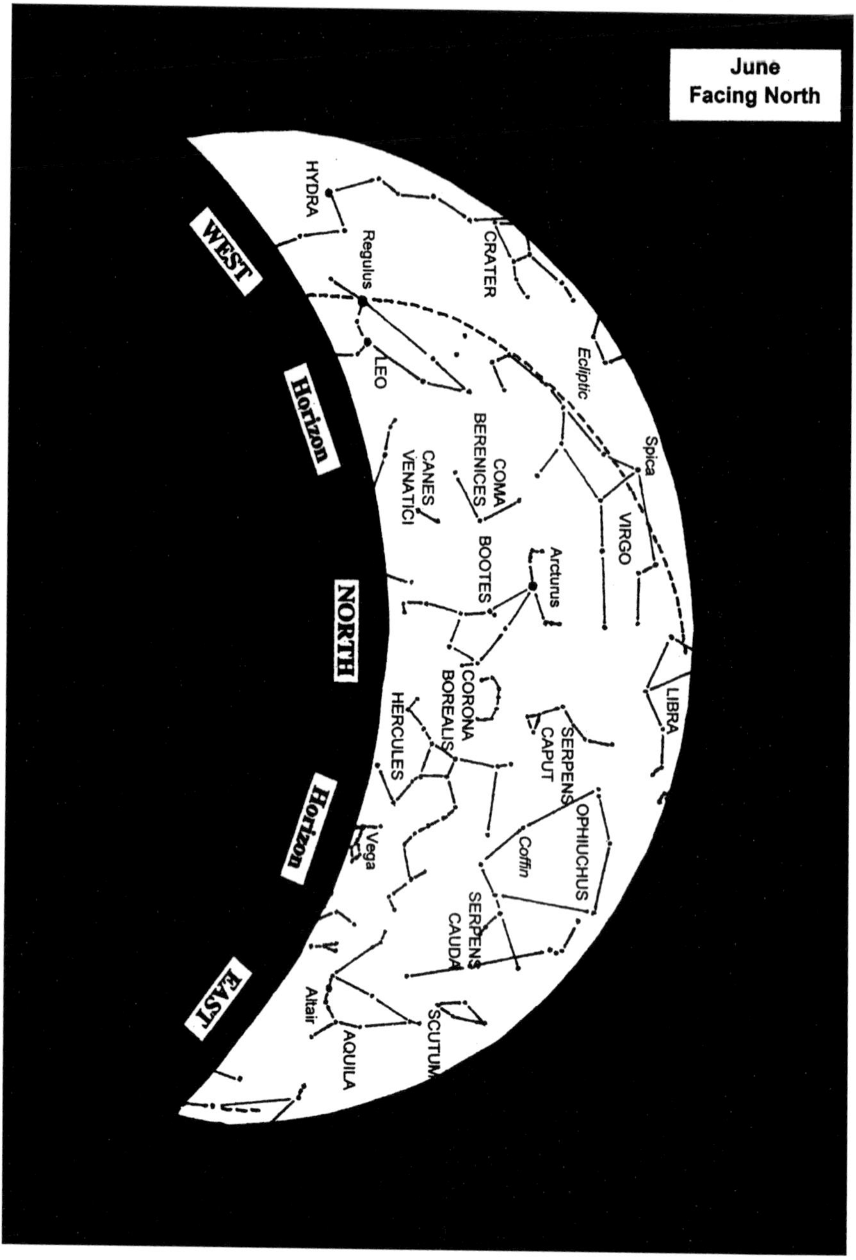
June
Facing North
HYDRA
WEST
Regulus
CRATER
LEO
Horizon
Ecliptic
COMA
BERENICES
CANES
VENATICI
Spica
VIRGO
Arcturus
BOOTES
NORTH
CORONA
BOREALIS
HERCULES
LIBRA
SERPENS
CAPUT
OPHIUCHUS
Coffin
Vega
Horizon
SERPENS
CAUDA
EAST
Altair
AQUILA
SCUTUM

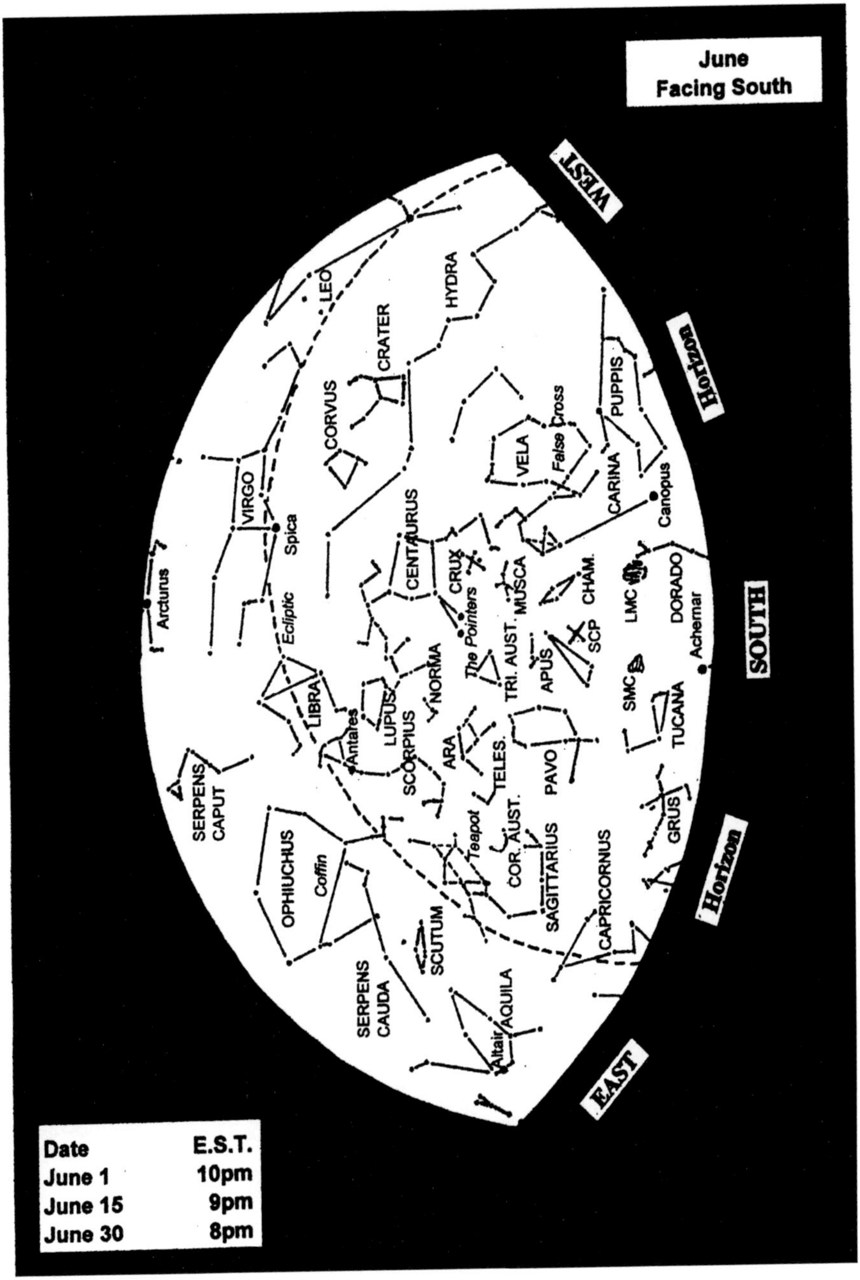
June
Facing South
WEST
Horizon
SOUTH
Horizon
EAST
Date E.S.T.
June 1 10pm
June 15 9pm
June 30 8pm
LEO
HYDRA
CRATER
CORVUS
PUPPIS
False Cross
VELA
CARINA
Canopus
VIRGO
Spica
CENTAURUS
Arcturus
Ecliptic
CRUX
MUSCA
CHAM.
LMC
DORADO
The Pointers
TRI. AUST.
SCP
Achernar
NORMA
APUS
SMC
LIBRA
LUPUS
TUCANA
Antares
SCORPIUS
ARA
TELES.
PAVO
SERPENS
CAPUT
GRUS
Teapot
COR. AUST.
SAGITTARIUS
OPHIUCHUS
Coffin
CAPRICORNUS
SCUTUM
SERPENS
CAUDA
AQUILA
Altair

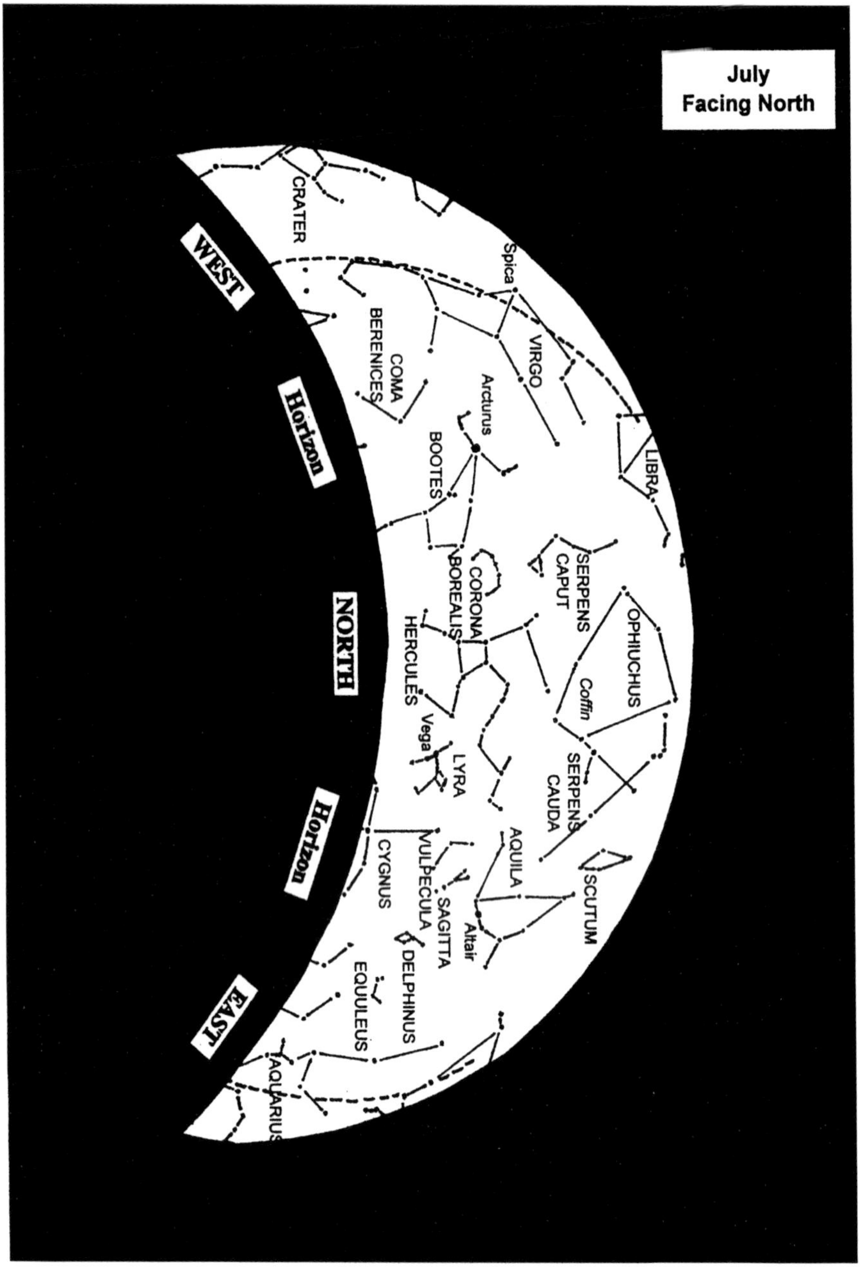
July
Facing North
WEST
Horizon
NORTH
Horizon
EAST
CRATER
Spica
VIRGO
COMA
BERENICES
Arcturus
BOOTES
LIBRA
SERPENS
CAPUT
CORONA
BOREALIS
HERCULES
OPHIUCHUS
Coffin
Vega
LYRA
SERPENS
CAUDA
AQUILA
SCUTUM
CYGNUS
VULPECULA
SAGITTA
Altair
DELPHINUS
EQUULEUS
AQUARIUS

July
Facing South

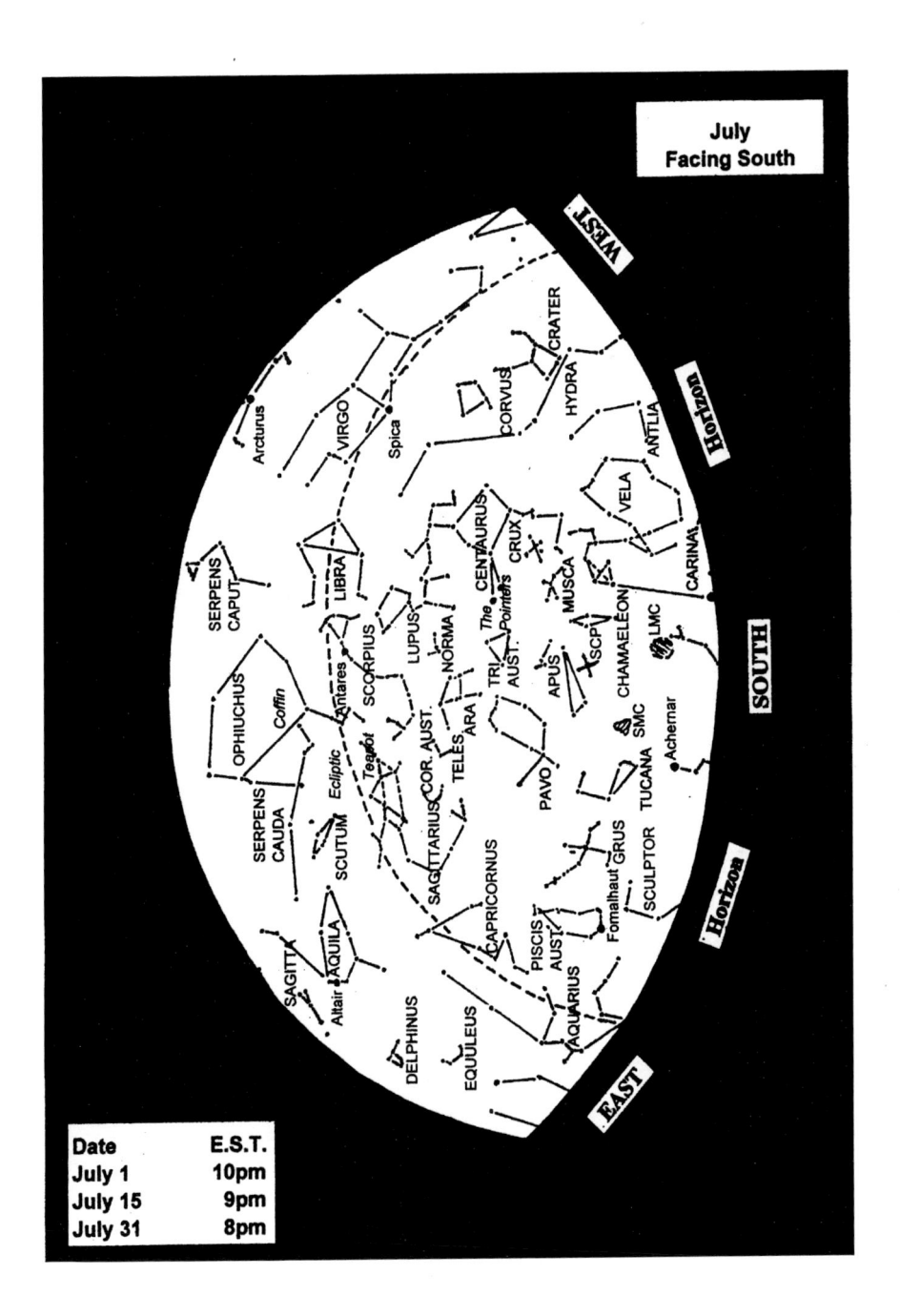

Date	E.S.T.
July 1	10pm
July 15	9pm
July 31	8pm

72

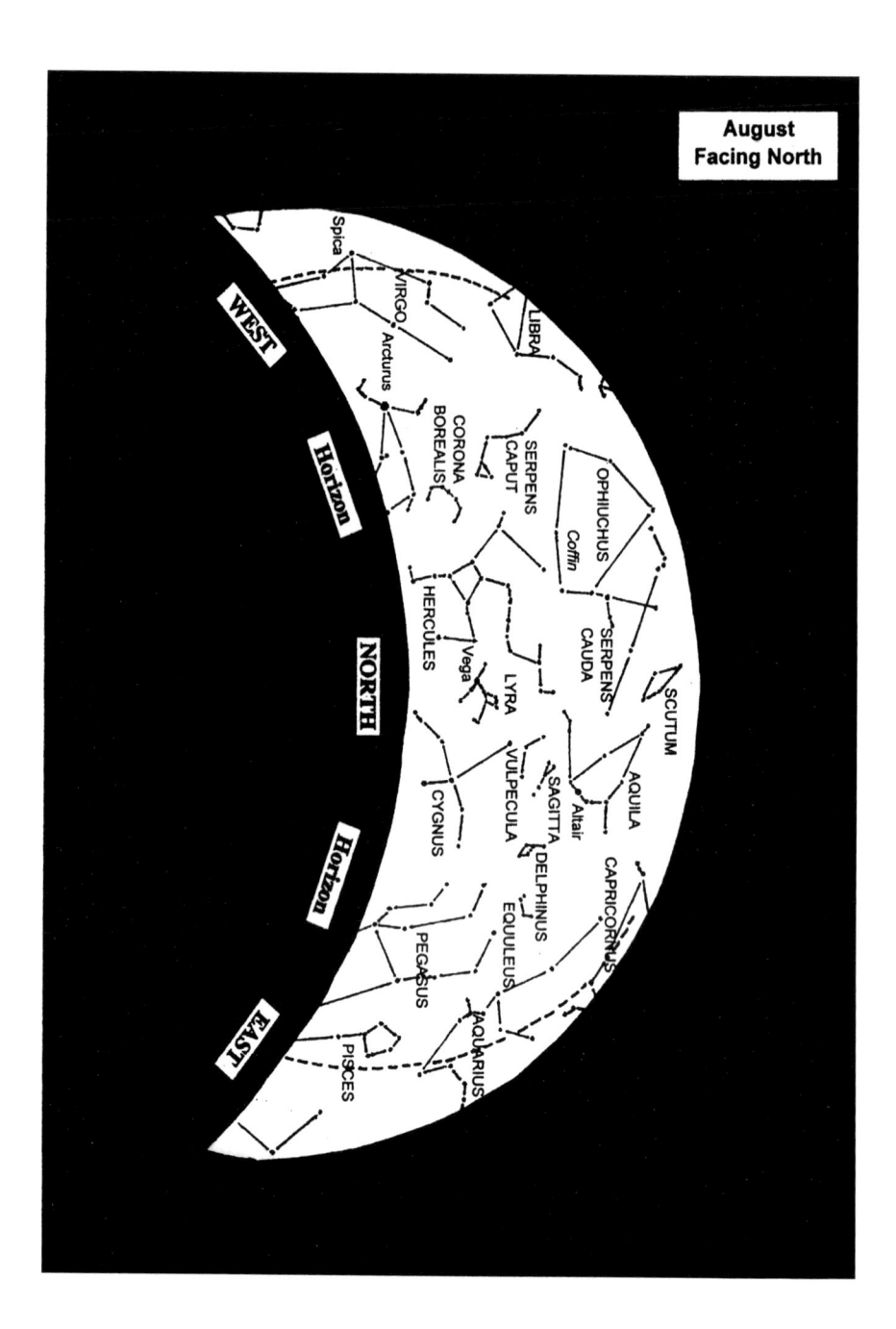

**August
Facing South**

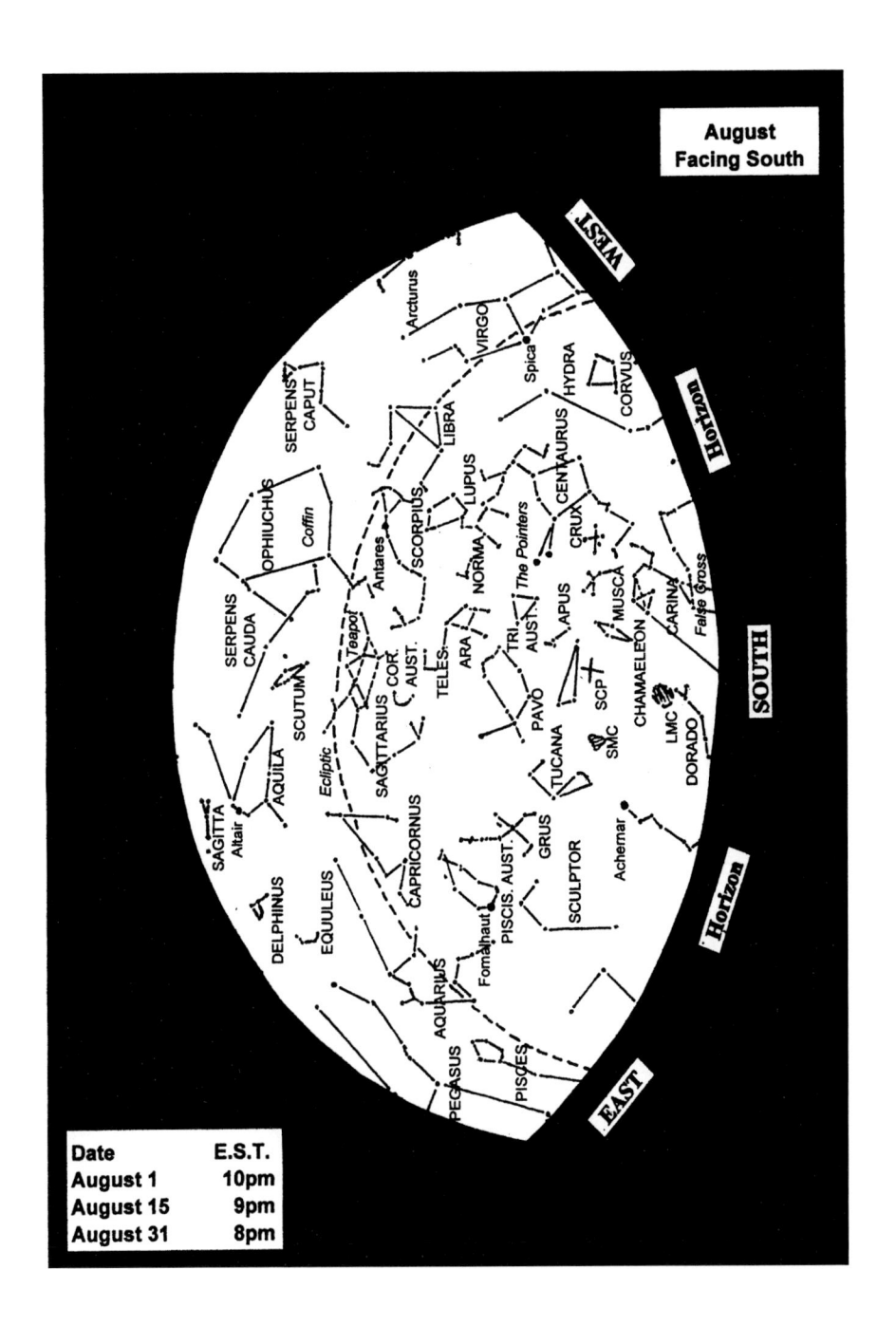

Date	E.S.T.
August 1	10pm
August 15	9pm
August 31	8pm

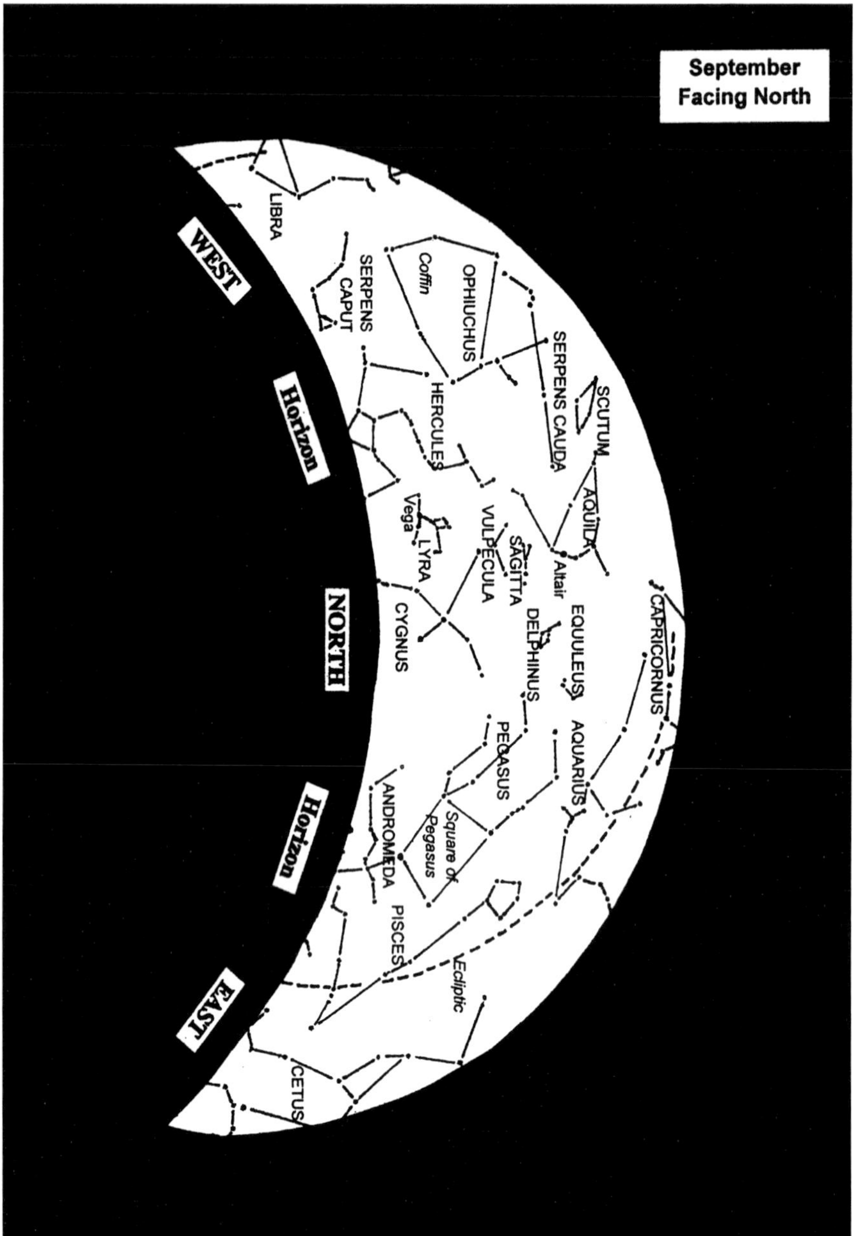
September
Facing North
WEST
Horizon
NORTH
Horizon
EAST
LIBRA
SERPENS
CAPUT
Coffin
OPHIUCHUS
SERPENS CAUDA
SCUTUM
HERCULES
AQUILA
Altair
Vega
LYRA
VULPECULA
SAGITTA
CYGNUS
DELPHINUS
EQUULEUS
CAPRICORNUS
AQUARIUS
PEGASUS
Square of
Pegasus
ANDROMEDA
PISCES
Ecliptic
CETUS

75

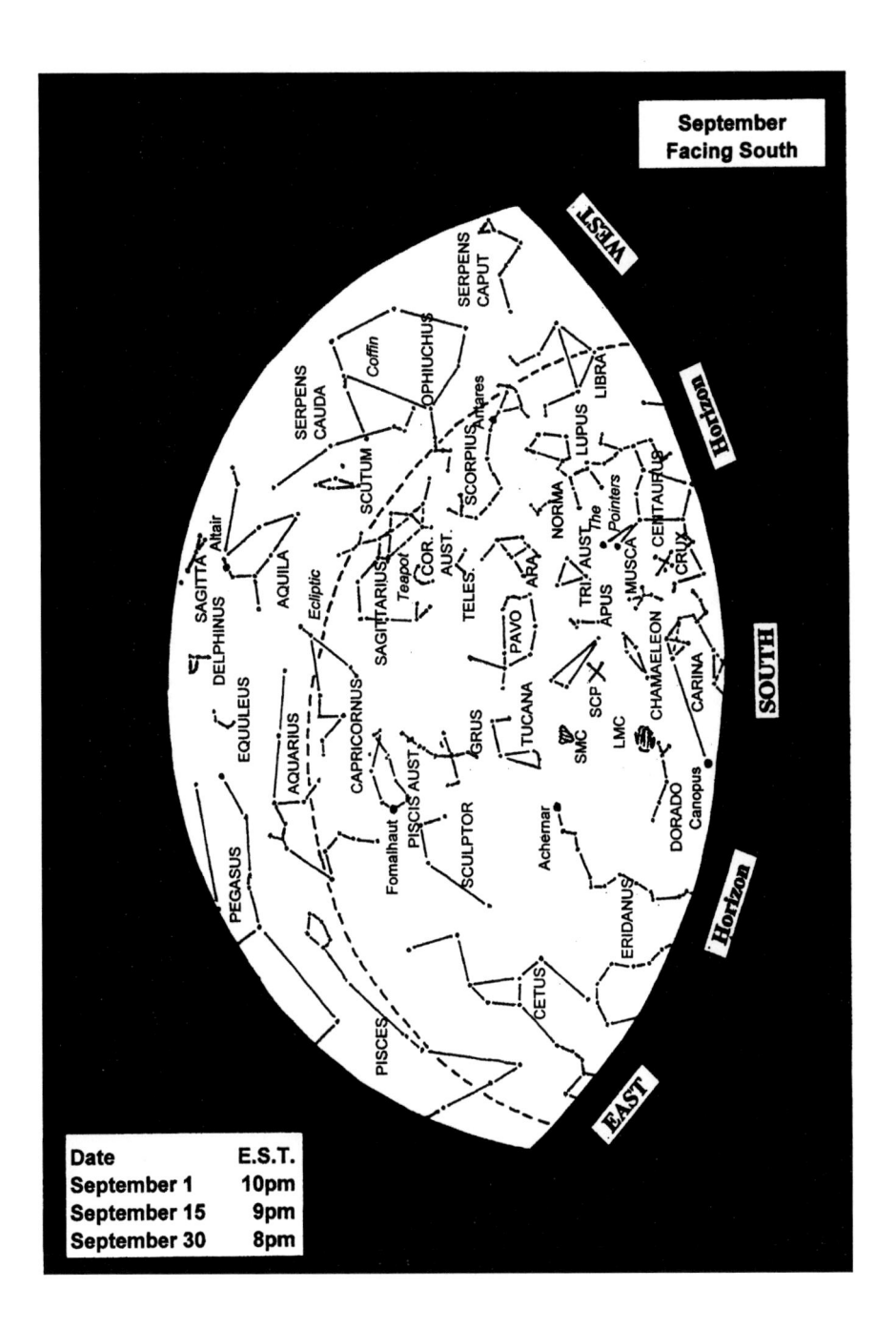

October
Facing North

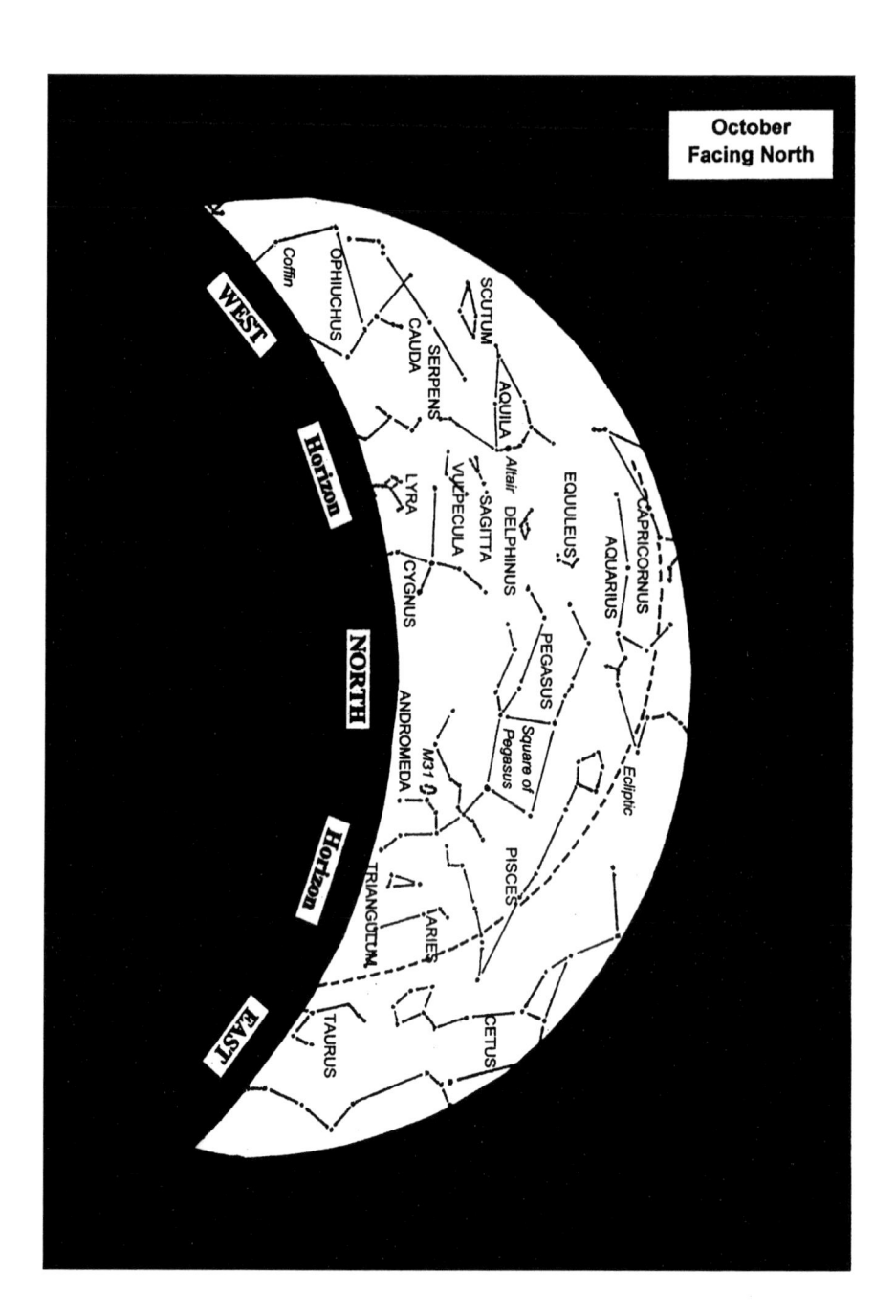

**October
Facing South**

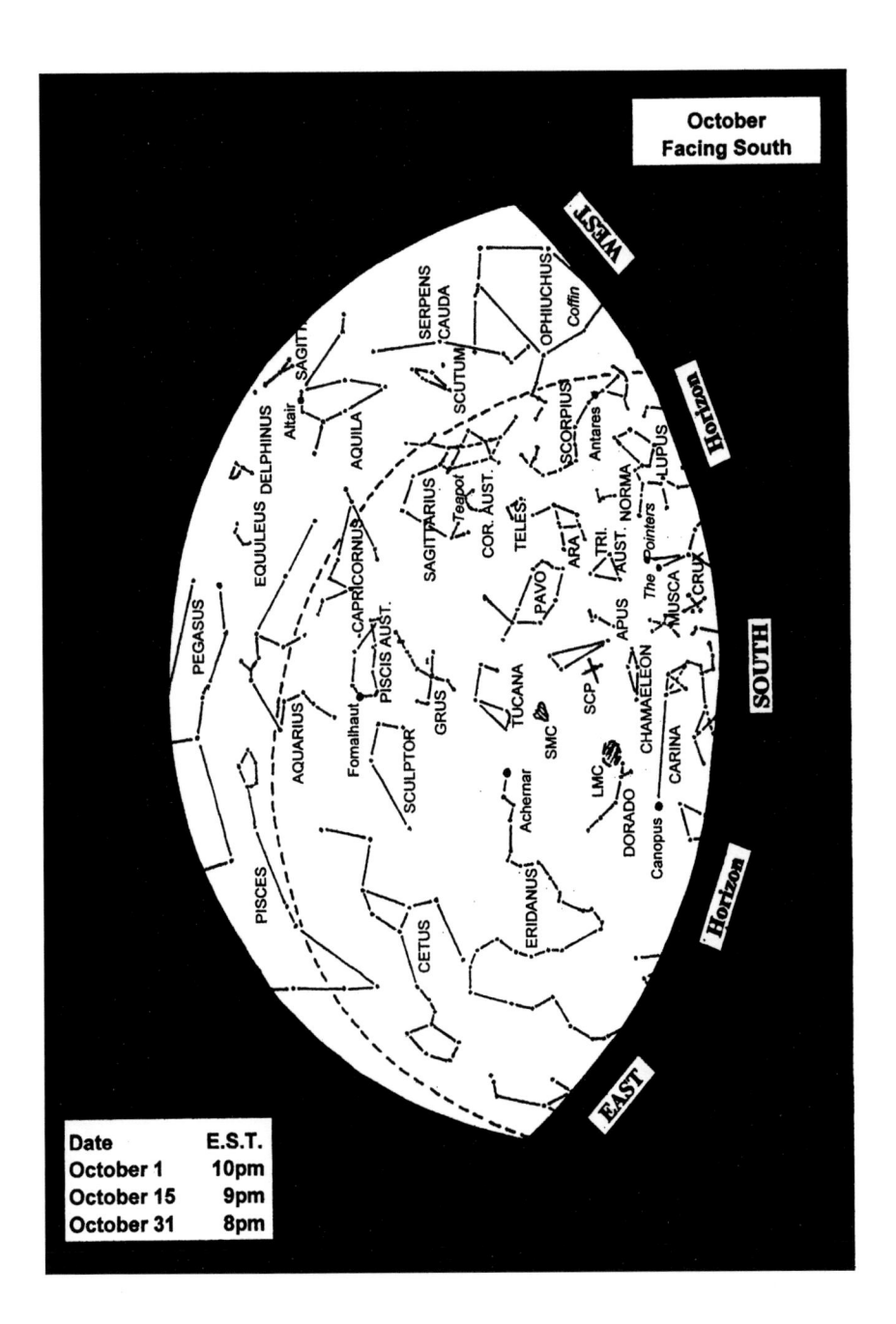

Date	E.S.T.
October 1	10pm
October 15	9pm
October 31	8pm

November Facing North

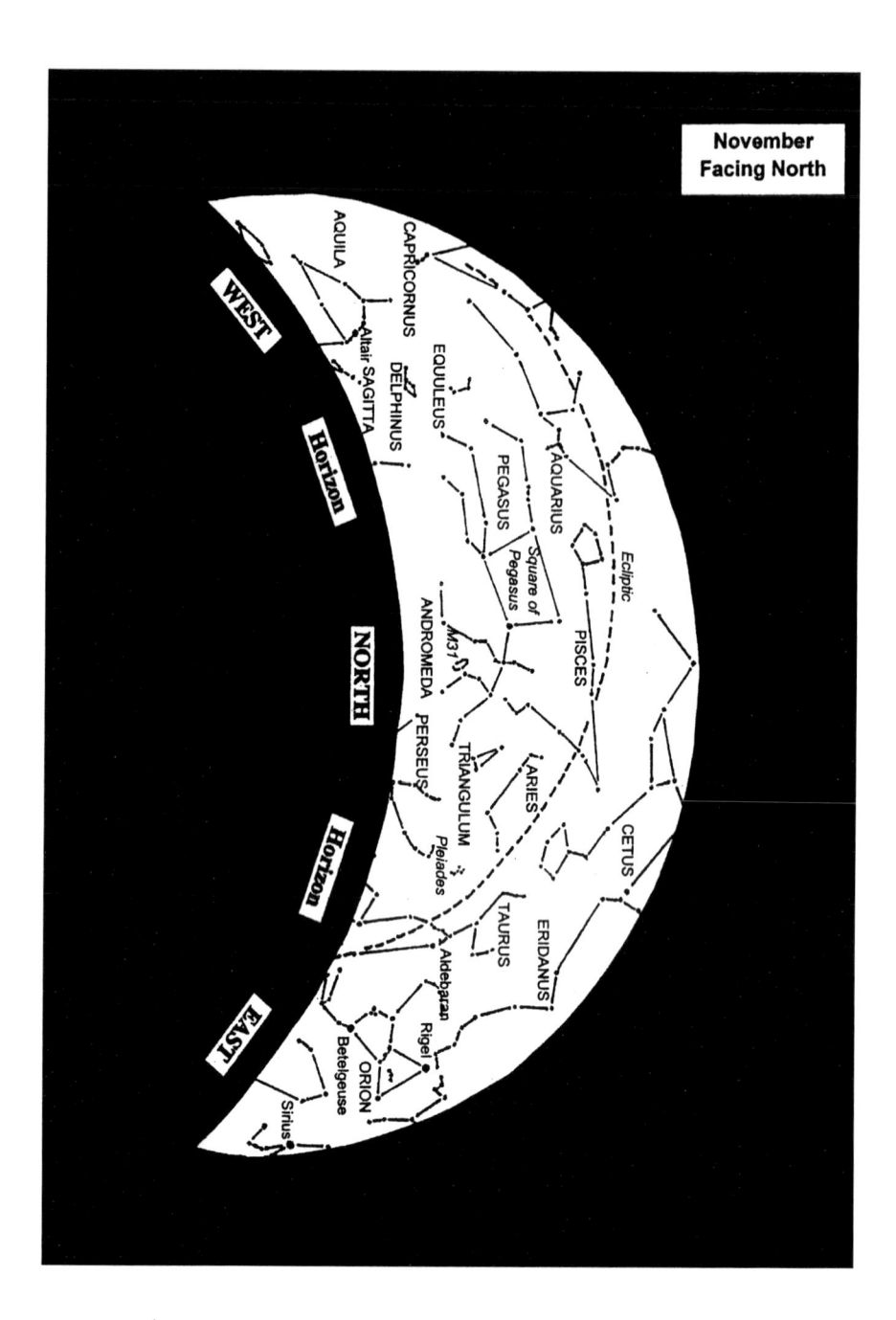

November Facing South

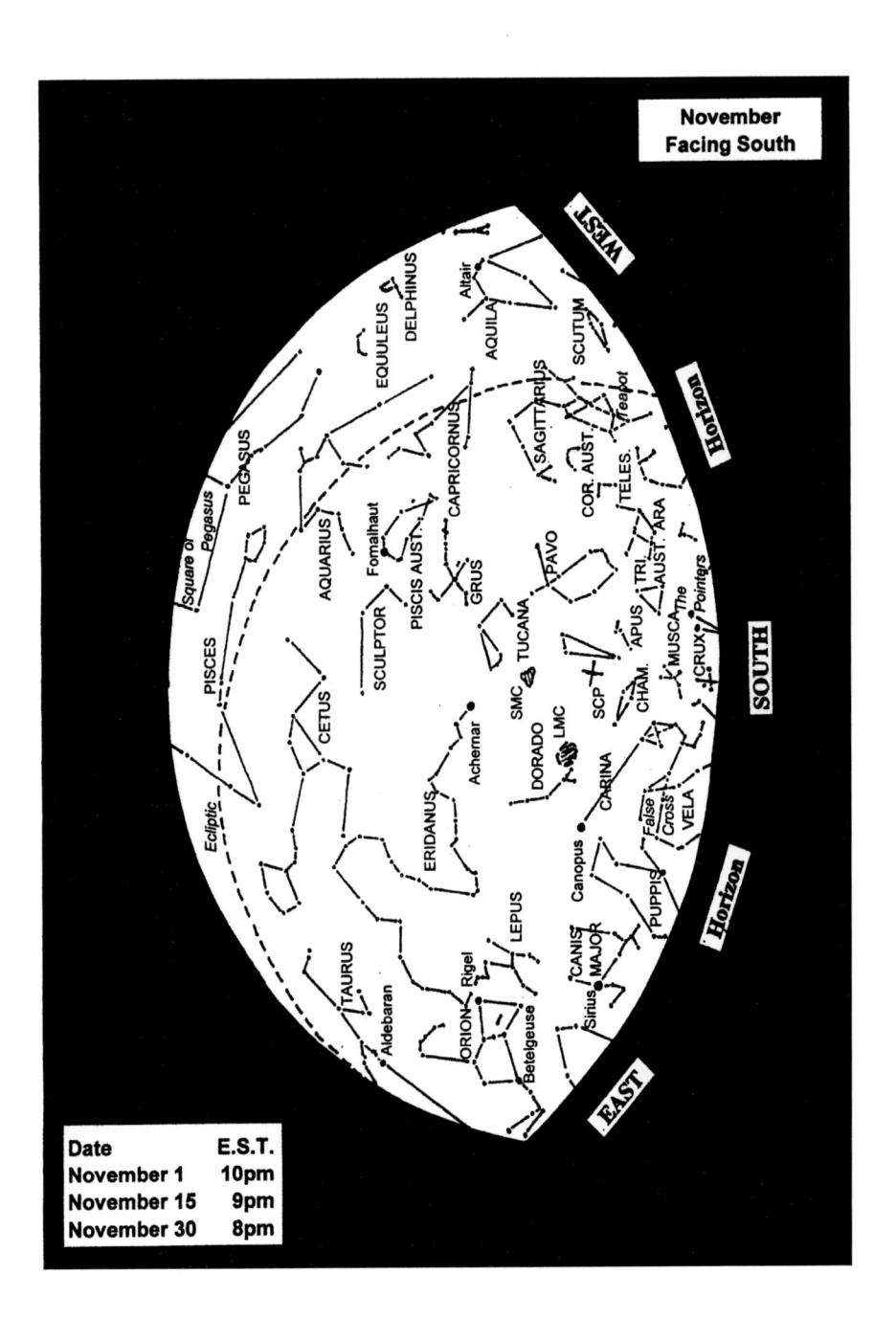

Date	E.S.T.
November 1	10pm
November 15	9pm
November 30	8pm

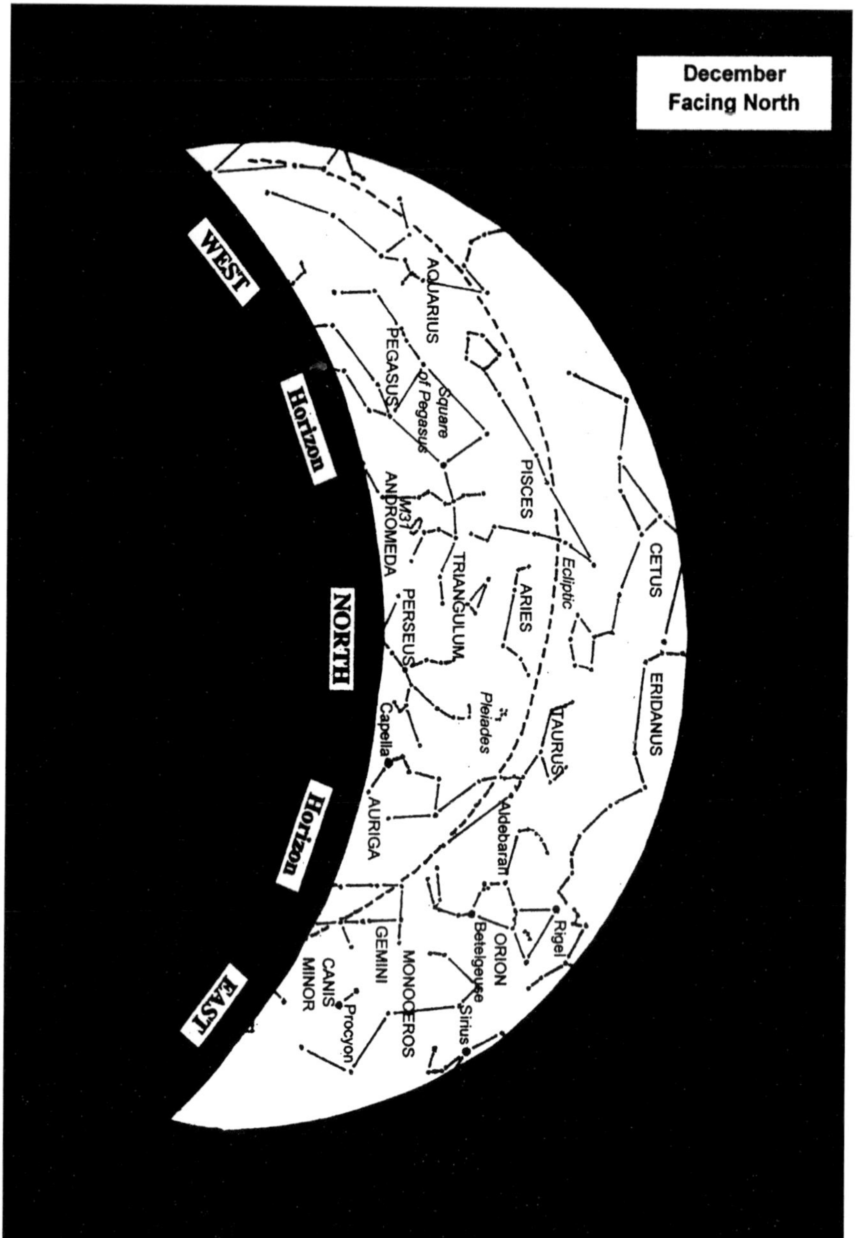
December
Facing North
WEST
Horizon
NORTH
Horizon
EAST
AQUARIUS
PEGASUS
Square of Pegasus
PISCES
ANDROMEDA
M31
TRIANGULUM
ARIES
Ecliptic
CETUS
PERSEUS
Pleiades
TAURUS
ERIDANUS
Capella
AURIGA
Aldebaran
Rigel
ORION
Betelgeuse
GEMINI
MONOCEROS
CANIS MINOR
Procyon
Sirius

December
Facing South

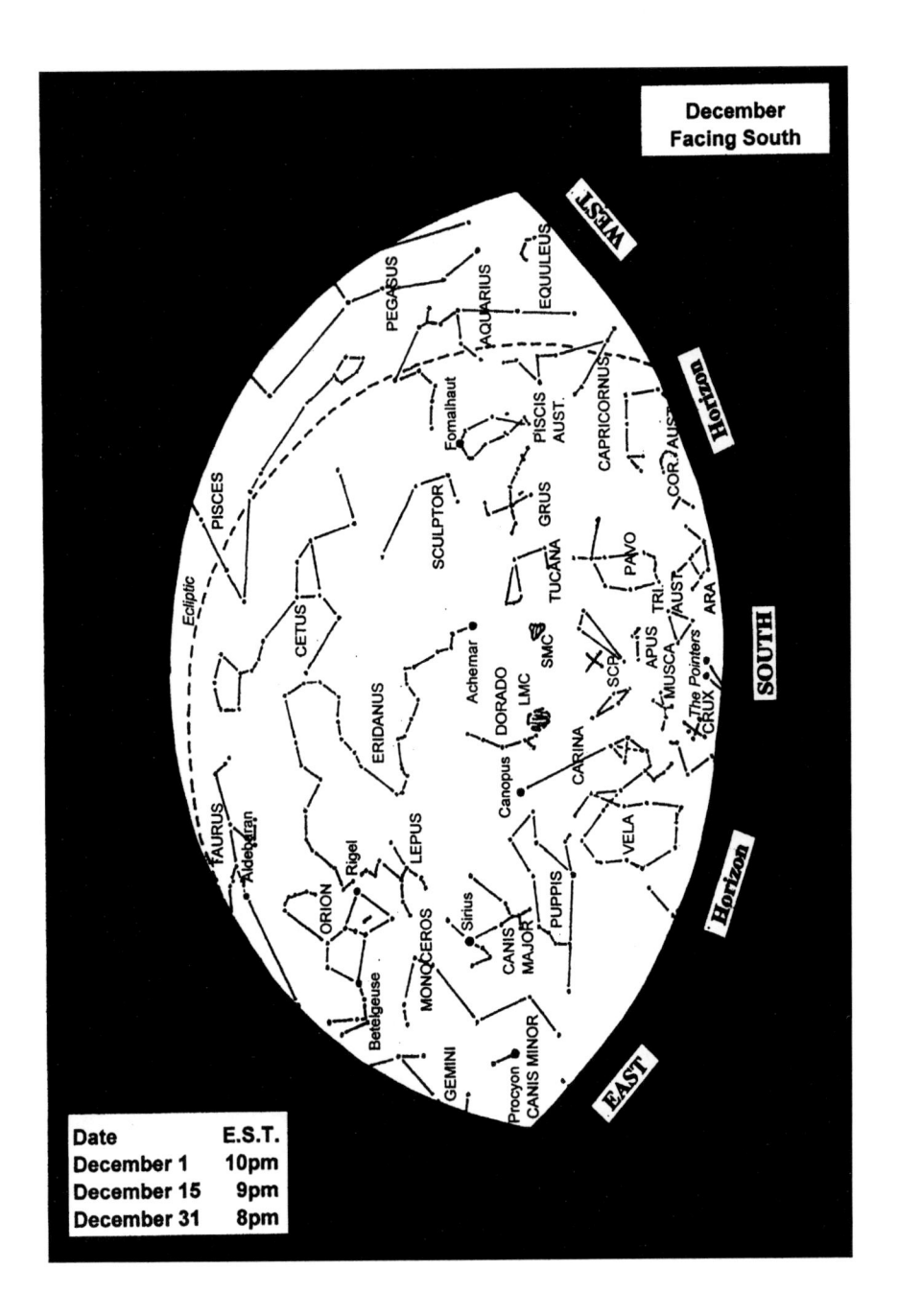

Date	E.S.T.
December 1	10pm
December 15	9pm
December 31	8pm

Andromeda (October to December)
(Northerly aspect)

Andromeda is situated low over the northern horizon. The higher your latitude, the better will be your view of it. But, even as far south as Sydney, you can see it *if* you can escape the glare of the city's light pollution and have an unobstructed view of the northern horizon. The main claim to fame of this constellation is the famous Andromeda Galaxy.

This constellation is named after the daughter of the vain Queen Cassiopeia and King Cepheus. This lass has the dubious privilege of being chained to a rock as a sacrifice to the sea monster Cetus, but was saved by the hero Perseus. All very dramatic but to us, it looks like a hockey player with his stick. It is most easily located by first finding the prominent Square of Pegasus (who Perseus was riding when he rescued Andromeda). The alpha star of Andromeda is shared with the bottom right star of the Square.

First find the long narrow triangle in Triangulum. Just over 3° below and west of the bottom left star you will find **56 Andromedae** – an optical double star. Binoculars will easily split this pair comprising a yellow giant star, magnitude 5.7 and an orange giant of magnitude 5.9.

Though they appear a natural pair, they are in fact 320 l.y. and 990 l.y. away respectively. Try the 'defocusing' trick to see the difference between the yellow and the orange.

Now find α And. at the bottom right corner of the Square of Pegasus. Move 7° east to δ And. and then 8° north-east to β And. Then move down NW about 7° following the two fainter stars. Just beyond the second star you should see **M31 – the Andromeda Galaxy**. This is something really special – what a way to start your tour. It is the furthest thing your naked eye will ever see – a whopping 2.4 million light years away. M31, also called **NGC224**, is a great spiral galaxy, the nearest spiral to our own. (The Magellanic Clouds are much closer but are irregular galaxies, not spirals. See Dorado and Tucana.) This is what our Milky Way would look like from 2 million light years away. Though visible in the naked eye as a hazy egg shape, it is best seen through binoculars because of their low magnification and light collection. **M31** has a diameter of about 150,000 light years and contains at least 400 billion stars. ***Wow!***

Antlia – the Air Pump (March to June)
(High overhead)

Not one of your more romantically named constellations nor dramatically shaped figures. This is typical of most latter discovered southern hemisphere constellations – the ancient astronomers and name givers didn't see them, but down to earth (or down to sea) explorer-navigators did.

Antlia's stars are very faint and the constellation so insignificant, it often does not appear on star maps. It is easiest found high in the sky about two False Cross lengths directly above the False Cross, on the opposite side of Vela (the Sails).

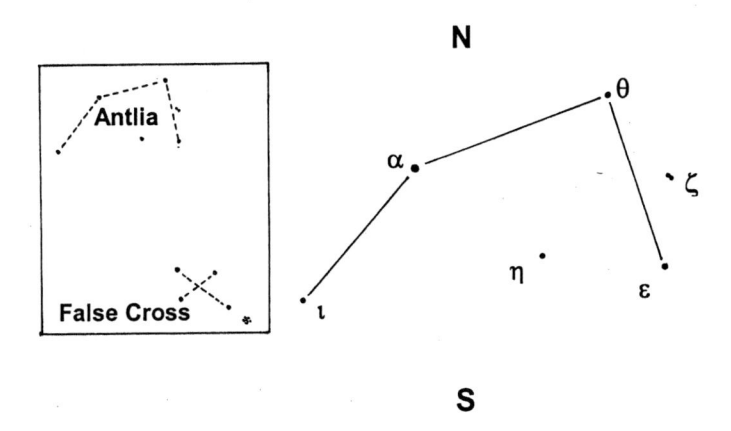

ζ (Zeta) Antliae is a wide optical double, barely visible to the naked eye. Its component stars are ζ^1, a blue-white of magnitude 5.8 and ζ^2, magnitude 5.9. Though your binoculars won't show it, ζ^1 is itself a true binary star.

What is a Red Giant?

As part of a star's life cycle, when a star ages and uses up most of its hydrogen fuel in its core, through a complicated process, it swells in size to huge dimensions. For example, when our Sun does this in around 4 billion years, its bloated red sphere would almost fill our sky. As it does its outer surface cools and becomes red.

Stars more massive than our Sun (such as Antares and Betelgeuse) grow so large they would engulf all the planets out to Mars. These are red supergiants.

Apus – The Bird of Paradise (All year)
(Southerly aspect)

This constellation, though faint and unremarkable, is very close to the South Celestial Pole (SCP) and is viewable all year, though it is more conveniently higher from April to September. It was named in honour of that magnificent bird of plumage from Papua New Guinea. Find the SCP and move about one third the distance towards α Centauri to find Apus.

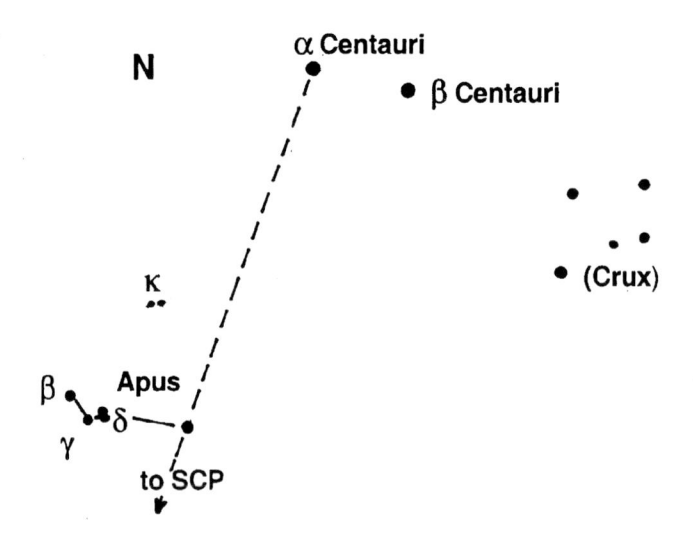

δ **(Delta) Apodus** can be seen 5° to the left of the line between the SCP and α Centaurus as an optical double. Look for the narrow triangle made up of β, γ and δ. δ is the most western of the three. While appearing a single star to the naked eye (the pair are only 1' apart), it shows up beautifully as a double in binoculars. The pair are both orange giant stars of magnitude 4.7 and 5.3. The fainter star is 760 light years away and the brighter 100 light years beyond it.

κ **(Kappa) Apodus** is a wide naked eye optical double, referred to as κ1 and κ2. They are about ½ ° apart (a Moon diameter) and are both blue-white stars about magnitude 5.5.

Aquarius – The Water Carrier (August to November)
(A northerly aspect)

One of the well known constellations named by the ancients. The complex shape representing a figure pouring water (or wine) from a jar for the gods is best recognised from the distinctive 'Y' shaped group of stars at its centre. (The Y is approx. 4° long from side to side.) This is the water jar. Mythology says the figure is a shepherd boy named Ganymede who became the wine bearer for Zeus and the Olympus gods and ultimately had the largest moon of Jupiter (and the solar system) named after him. This shows that it always pays to please the boss.

Aquarius has a couple of distinctive binocular objects.

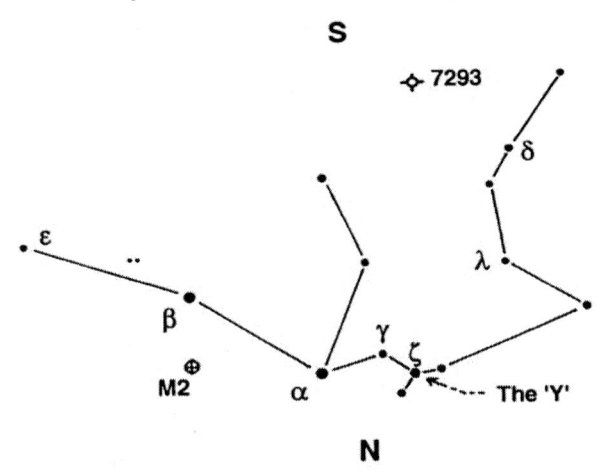

First, follow the line from ζ (the centre of the Y) to α (the bright star to its west), extend the line past α by 1.5 times the distance between ζ and α, and there in your binoculars you will find **M2**, a very nice concentrated globular cluster about 37,000 light years away. While just out of naked eye visibility at magnitude 6.5, it is a great target for binoculars. Then, 20° above (south of) the 'Y' and almost directly overhead, you will find a faint circular foggy patch, **NGC7293**, a very famous planetary nebula called the **Helix Nebula**.

Planetary nebulae are the result of a Sun-sized star reaching the end of its long life. The star cools and swells to a huge size, then sheds its outer layers of material, casting it out into space as an ever expanding shell. The remainder of the star contracts into a white dwarf, an Earth sized star of incredibly high density. The expanding shell appears as a circular ring, like a smoke ring or a doughnut. Sometimes, very odd shapes result.

The Helix Nebula is a spectacular example of this phenomenon, possible because of its close proximity to us (only 300 light years away). If you could see all of it, it would appear as a smoke ring about half the apparent diameter of the Moon. Remember, this is how our Sun will end in about 5 billion years.

Aquila – the Eagle (July to October)
(A northerly aspect)

An ancient constellation representing the bird that caddied for Zeus by carrying his thunderbolts for him. It also did other dastardly deeds for Zeus, the best known being the perpetual torment of the Titan Prometheus by returning day after day to eat his liver until Hercules dispatched the eagle with an arrow (Sagitta). Though, unfortunately there are no binocular objects in Aquila, it is included here because the distinctive shape of the line of three stars, pointing down at about 60°, brightest star in the middle, makes it easy to spot and acts as a launching pad for finding other constellations such as Capricornus, Sagitta and Vulpecula.

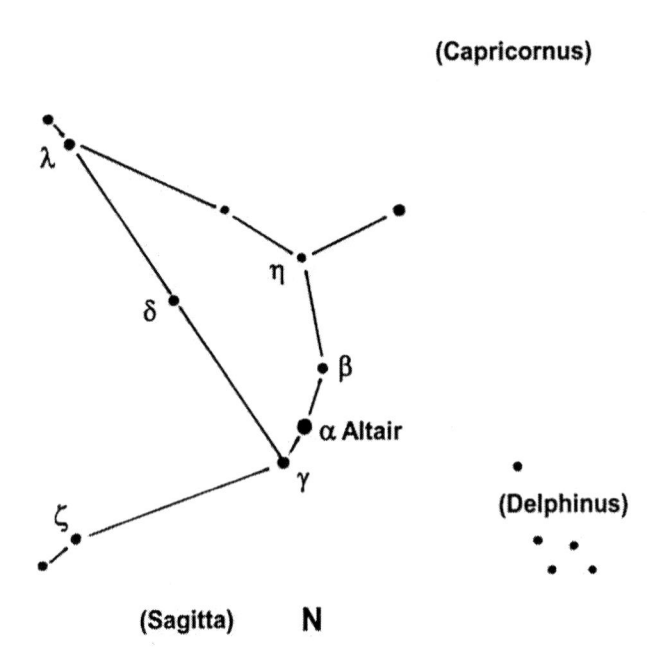

α **(Alpha) Aquilae** is the 12th brightest star in the sky and is also amongst the closest, at only 17 light years. It has the name **'Altair'** (meaning 'the flying eagle') and is of magnitude 0.77. It can be easily spotted as the bright centre star of the three. For sci-fi buffs, Altair is famous as the star orbited by the 'Forbidden Planet', the movie that brought you Robbie the Robot and 'the monster from the id'.

Ara – the Altar (April to October)
(Southerly aspect)
(Circumpolar but very low between November and March.)

Not a well known constellation, it was named by the Greeks for an altar used by the gods of Olympus as they prepared to go into battle with the Titans. It is easiest found just beyond the curl of Scorpius' tail, in the direction of the South Celestial Pole (SCP).

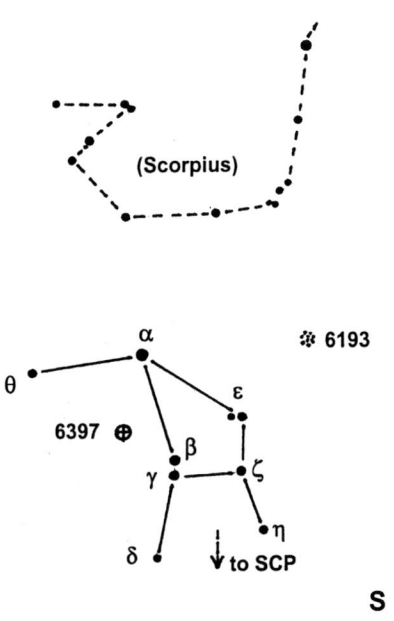

First an easy double star. ε **(Epsilon) Arae** is a naked eye optical double seen to the north-west of the bright stars β and γ. Though brighter (mag. 4.0), the reddish ε^1 is in fact 200 light years further away than the 67 light year distant ε^2, a yellow mag. 5.4 star.

Now find mid-point between α and γ Arae, then move about 2° east. You will find **NGC6397,** one of the closest globular clusters to Earth (a 'mere' 10,500 light years away). As a result, it is nicely viewable in binoculars, though it appears, at 6th magnitude, barely visible to the naked eye except in extra dark skies. Use the chart above to find it, it's easily done. This 'glob' has a concentrated centre and then the stars thin out towards its 'edge'.

Then if you move west of α Arae till you are level with Scorpius' spine, you will find **NGC6193**. This 5th magnitude open star cluster (that is, just visible to the naked eye) should be observable in your binoculars. Comprising about 30 stars of 6th magnitude and fainter, it is 4,200 light years away.

Auriga – the Charioteer (December to February)
(Very low in the north)

Another constellation named by the ancient Greeks for one of their legendary kings. You will need a clear northern horizon, free from light pollution.

Auriga contains a nice set of open clusters.

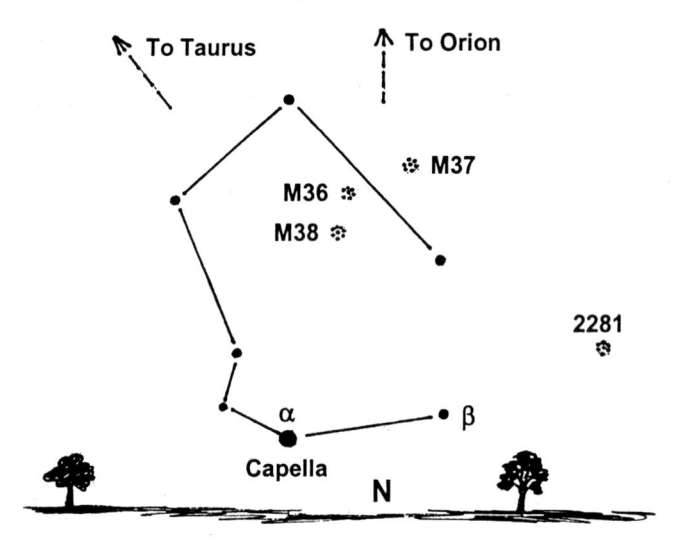

α **Aurigae** is Capella, the 6th brightest star in the sky, magnitude 0.08 and is found very low above the northern horizon. Though still 'only' a star in binoculars, it is interesting to know that in fact it is two giant yellow stars orbiting very close to each other every 104 days. Capella is 42 light years away. (*Trivia: If you happen to be 42 years old, the light you are seeing from Capella left that star the year you were born.*)

10° above Capella is **M38**, an open cluster 3,900 light years away. This cluster with about 100 stars is fairly large and dispersed and may appear just as a fuzzy patch. Then, about 2° above and to the right of M38 find **M36**, a small but bright 'knot of light' containing about 60 stars. This open cluster is 3,900 light years away also. Now move 3° to the right and about 1.5° up to find **M37**, a large open cluster 4,200 light years away. This one has about 150 stars in it and through binoculars looks like a hazy patch of light.

Finally, follow the line from Capella to β (a 2nd mag. white star) an equal distance plus a quarter to spot **NGC2281**, a smaller cluster, containing only 30 stars. However, as it is much closer (only 1,500 light years away) the individual stars can be seen (that is, resolved) through binoculars.

Bootes – the Herdsman (May to July)
(Low in the north)

Another ancient Greek constellation, based on mythology. The herdsman involved is seen either as leading hunting dogs (in the constellation Canes Venatici) or herding a bear (another constellation, Ursa Major, which can only be seen from the higher latitudes).

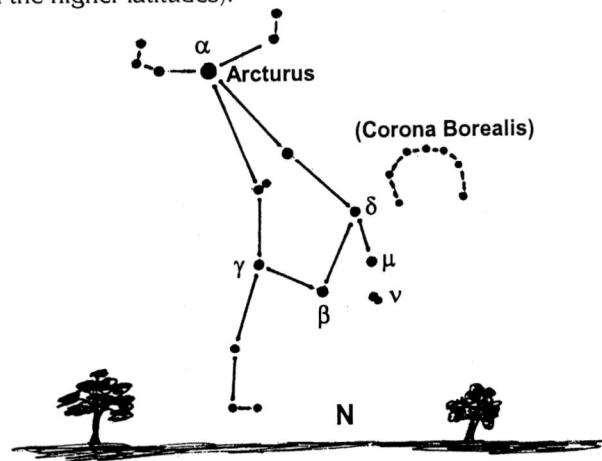

α **Bootis** is called **Arcturus,** meaning 'the Keeper of the Bear', a star that has many claims to fame. One is that it is the brightest star that lies to the north of the celestial equator. This is what helps to find Bootes so easily. From the monthly maps, you know roughly where Bootes should be found. You look there and see this very bright orange star in the northern sky. Its colour is very striking in binoculars. That's Arcturus. Overall, it is the 4th brightest star in the sky at magnitude –0.04. Arcturus is a relatively close neighbour of ours, at only 37 light years. It is an orange giant about 27 times the diameter of our Sun and about four times the mass. This makes it only about 3/10,000th times as dense as our Sun – not very dense. Of particular interest about Arcturus is that it is only 'passing through' the neighbourhood. In effect, it is part of a stream of stellar traffic that is orbiting the hub of our galaxy in a different plane to the stream our Sun is in. You might say it is cutting across our path like a car at a roundabout. In a few thousand years, it will have gone past us and head out in the opposite direction.

The next stars are even closer to the northern horizon, 'below' Arcturus. 15° below and about 12° east of Arcturus is the 3.5 mag. star δ **Bootis,** an optical double star that stands out well in binoculars. However, unlike normal 'optical doubles' it is not merely an optical illusion of very distant stars being in line and appearing to be close together. They are what is called a 'common proper motion pair'.

That means they are close together in space and are travelling along together but do not orbit each other. Effectively, they are 'common travellers'.

What looks to the naked eye as a magnitude 3.5 yellow star reveals in binoculars a giant yellow star with a fainter companion 7th magnitude yellow star (just like our Sun) with a relatively wide separation. In fact, they are about 0.7 light years apart. The pair are about 117 light years away.

From δ, drop down another 4° and go 2° east to find μ **Bootis**, another 'common proper motion pair' with a twist. One of the pair is itself a genuine binary star. μ is about 1,200 light years away. Your naked eye will see a magnitude 4.3 blue-white star while binoculars will reveal the yellowish second star with magnitude 6.5 widely separate from the first. In fact they are about 0.4 light years apart. Imagine how bright one would seem to the other when you consider our closest star, α Centauri, is 10 times further away from the Sun. Though your binoculars will not reveal it, this fainter companion star is a binary star with two Sun-like stars orbiting each other every 260 years, similar to the time it takes Pluto to orbit the Sun.

Finally, drop down another 4° (getting close to the horizon) and 1° east to find ν **Bootis**, an optical double that shows up well in binoculars. Both stars, a white and an orange giant appear the same magnitude (5.0) but in fact are 430 and 870 light years distant respectively. That is, one is twice as far away as the other.

What is a White Dwarf?

After a star about our Sun's size reaches the red giant stage at the end of its life, it 'shucks off' its outer surface layer which is seen as a beautiful planetary nebula. What is left is the exposed central core of the star, depleted of hydrogen and helium which have fused into heavier elements, containing about 90% of the star's original mass. Due to gravity, the star contracts to a very small size, roughly that of Earth. This star is extremely dense and hot and glows white – a white dwarf. It is so dense that a tea-spoon full of it would weigh as much as a family car. However, due to their small size, there are none that are visible to the naked eye.

A white dwarf is no longer an active star – its furnace has turned off – so that it gradually cools off, but this can take billions of years. After then, it simply fades, and fades, and …

Unless it is part of a binary system. Then it is a completely different story. See the story on Type 1a supernovae.

Cancer – the Crab (February to April)
(A northerly aspect)

Though its name has an unpleasant connotation, all it means is 'crab'. This constellation consists of mostly faint stars (its brightest is magnitude 3.9) but it has a distinct 'Y' shape and using the monthly star maps, you will find it easily in a dark sky immediately to the west of Leo's nose. The thing to remember is its size – the 'Y' is about 3 times bigger than the Southern Cross.

Mythologically, the Crab is the one which nipped the heel of Hercules, who subsequently crushed it under said heel.

Cancer is one of the constellations in the Zodiac, though a very faint one. Despite its dubious astrological significance, it has a number of interesting astronomical objects for the binoculars.

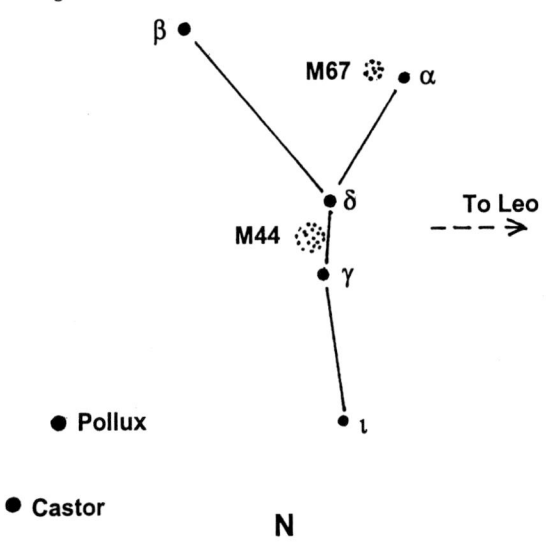

ι **(iota) Cancri,** the bottom star of the big Y, is a binary double consisting of a naked eye yellow giant (magnitude 4.0) and a blue-white companion of magnitude 6.6. They are 300 light years away. They are relatively close together and a test for your binoculars.

Now find δ Cancri, the centre of the Y. Move 2° down and 1° to the west of δ to find **M44,** one of those sky objects that show up best in binoculars. Its most common name is the Beehive Cluster (I like it), presumably because it resembles a swarm of bees. It has an official name of Praesepe, which means 'the manger'. This is supported by the names of the two stars below and above M44, γ Cancri and δ Cancri which are called Asellus Borealis (northern donkey) and Asellus Australis (southern donkey) respectively. The idea is that these stars represents two donkeys eating at the manger. M44 looks to the naked eye as a cloudy nebula, but binoculars reveal it to be a swarm of faint stars each of magnitudes greater than 6. On a good dark night, you should be able to see stars extending to an area as big as three Moon diameters. Galileo was the first to discover this back in 1610 with his puny but revolutionary telescope. It is estimated that there are over 200 stars in the related group, 570 light years away. It is one of the nearest of the galactic open star clusters.

A special occasion to look out for is the passage of a planet through the outskirts of M44. The ecliptic, the path of the planets across our sky, just passes M44 and it is a treat to pick out a planet with your binoculars as it dallies alongside the cluster. This is particularly the case with Jupiter as its four main moons, easily spotted in binoculars, seem to mingle with the faint M44 stars. Also, as δ Cancri is exactly on the ecliptic, it would not be unusual to see a planet 'collide' with it, or at least be seen 'in conjunction' with it.

Now find the top right star in the Y, α Cancri. 2° due west of this is **M67,** an open cluster far more remote than M44. M67 is 2,600 light years away, so it appears smaller and denser in your binoculars. Even so, in a dark sky it should appear as a misty oval about a large as the Moon. You won't be able to resolve individual stars. M67 has about 500 10th mag or fainter stars in it. It is a peculiar cluster, reputed to be one of the oldest known and actually located 2,500 light years *outside* the normal plane of our galaxy.

What is a Red Dwarf?

Not a BBC Sci-Fi TV show, but the name given to those stars that have masses much less than our Sun. Due to their low mass, they have been able to light up their nuclear furnace but not get hot enough to glow more than a dull red, about 3,000°C. Because they are of low mass, they are also much smaller in dimension and therefore difficult to see except with the largest telescopes.

Proxima Centauri, the companion star to Alpha Centauri, is a red dwarf.

Canes Venatici – the Hunting Dogs (May–June)

(Very low northerly aspect)

This supposedly represents two dogs being led in the hunt for the Great Bear. As a constellation shape, this leaves a lot to be desired, with very few distinctive stars. Its two brightest stars α and β are only mags. 2.8 and 4.2. It is best found by identifying Arcturus in Bootes and working from there with the monthly star maps.

However, it does contain some very interesting galaxies and clusters. These are listed here by way of a challenge. Being close to the limit of visibility in binoculars, they are not easy to see and will need a clear dark sky and a northerly horizon with no light pollution. However, if you do manage to spot them, you can take a large amount of pride in the achievement.

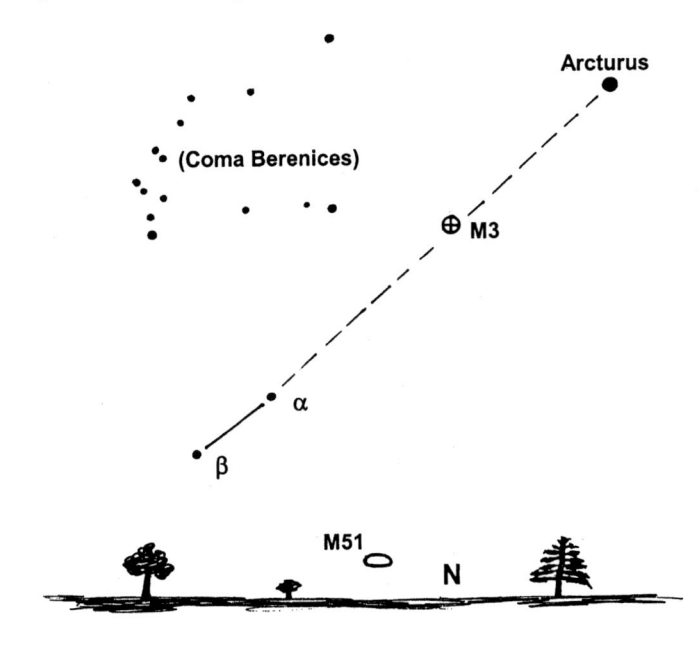

M3, if seen up closer, would be a most beautiful globular cluster. In fact it is – in a large telescope. However, it is along way off at 32,000 light years and is barely visible as a 6th magnitude star to the naked eye. In binoculars, it will be easily spotted and will appear as a fuzzy star. (It will not be as obvious as a globular as others are.) To find M3, locate Arcturus (in Bootes). Move about half-way to α CVn. and, voila, there it is.

Finding **M51** is trickier as it will be much closer to the horizon. First locate M3, then drop straight (north) another 20° and 3° to the west. M51 is an 8th magnitude object, on the edge of visibility with binoculars, realistically requiring larger binoculars. A challenge to see. You'll need a dark sky, no light pollution, a clear northern horizon and the use of averted vision.

It is a glorious spiral galaxy, NGC5194, about 20 million light years away. It bears the nick-name 'The Whirlpool Galaxy' and when you see a full colour photo, you'll know why. As I have indicated in an earlier section, do not expect to have your socks knocked off by what you see (an elongated hazy patch), but have the satisfaction of knowing you have glimpsed the light of hundreds of billions of stars that has travelled for 20 odd million years.

What causes a supernova?

In simple terms, a supernova is when a star, towards the end of its life, explodes catastrophically and can be seen from great distances, momentarily brighter than all the stars in its home galaxy.

Supernovae are hunted by some amateur astronomers as others hunt for comets. In far distant galaxies, they appear as a 'new' star against the familiar star pattern around the galaxy's fuzzy blob.

There are two basic types of supernovae. The best known is the type 2, which occurs when a massive star, greater than 3 times the mass of our Sun, uses up its core supply of hydrogen and it goes through a rapid process of fusing its elements into heavier and heavier elements. Ultimately, the core furnace goes out, gravity wins over radiation pressure, and the star collapses inwards almost instantly. There is a huge explosion and the star's material is hurtled into space and the star's core is left behind, either as a super-dense neutron star, or if big enough, a black hole.

The other type of supernova, a Type 1a, is covered elsewhere in this book.

Canis Major – The Greater Dog (December to April)
(High northerly aspect – almost overhead)

Canis Major is a constellation made up of many bright stars, the first five being of 1st and 2nd magnitude. It represents the larger of the two dogs that followed after Orion, the Hunter. When it is up, Canis Major is very easy to locate – just look for the brightest star in the sky. That is Sirius, the brightest star in both the southern and northern hemisphere. From the map below, you can see the dog's shape, with Sirius at the base of its neck. (Rotate the map 90° anticlockwise to see it better.) It might take you a bit longer to work it out in the sky without the lines joining the stars.

Canis Major is a bit tricky because its stars are not numbered (in the Greek alphabet) in order of their brightness. In order of brightness, they are: α (Sirius, mag. −1.47), ε (mag. 1.5), δ (mag. 1.8) and β(mag. 2.0). There would be some historical reason for this.

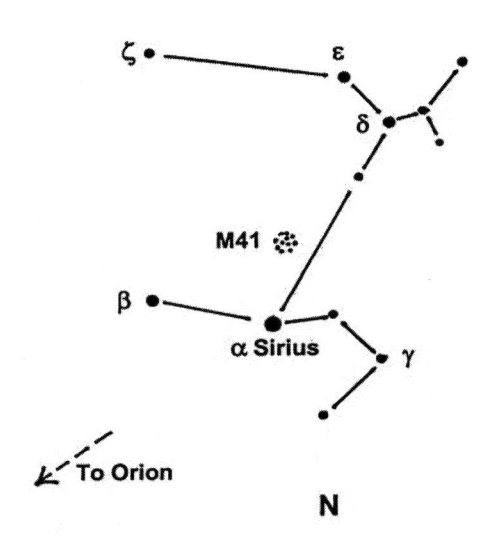

α **Canis Majoris (Sirius)** is also called 'The Dog Star', though its Greek meaning is 'scorching' or 'the sparkling one'. It is certainly brilliant to observe in binoculars or a small telescope. Though Sirius is intrinsically brighter than our Sun, its extreme brightness has more to do with its proximity, being the 5th closest star only 8.6 light years away (twice the distance of the nearest star, Alpha Centauri). If it was the same distance as α Centauri, it would be four times brighter than we see it now, making it approximately magnitude −3.0, very bright.

Sirius is actually a binary star, with a white dwarf companion. However, this is very difficult to spot with a telescope, impossible with binoculars. But it's there.

4° south of Sirius, just off the line (west) between Sirius and the dog's hind quarter, is **M41,** a very attractive open cluster 2,100 light years away. It is visible to the naked eye as a patch of light with overall magnitude of 4.5. With binoculars (on a good night) you should be able to resolve individual stars. It contains over 100 stars in the group, the brightest of which you should see in your binoculars. With a stand or a steady hand, you should be able to discern groupings or 'strings' of stars, including a 7th magnitude orange giant near its centre. Believe it or not, the Greeks were aware of this cluster circa 320B.C. referring to it as a 'cloudy spot'.

Canis Minor – This constellation does not receive a specific section in this book as it has no binocular objects of interest. However, its main star, Procyon, is a bright 0.34 mag. white star just east of halfway between Sirius and the twins Castor and Pollux (see Gemini). Procyon is only 11.4 light years away, a close neighbour (the 13th closest) to our Sun. And no reference to the Greater Dog would be complete without a mention of its companion, the Lesser Dog. My personal impression of the constellation Canis Minor is that it looks more like a dog's bone than a dog.

> ### A Type 1a Supernova.
> Subrahmanyan Chandrasekhar (1910 –1995) developed the theory of white dwarf stars and showed that if a white dwarf could collect additional material to increase its mass to 1.44 solar masses, it would explode as a supernova. If a white dwarf star happens to be part of a binary star system and if the other star is a largish one, interesting things can happen. As the companion stars goes into the red giant phase and bloats, its surface material can spiral across the gap and build up the white dwarf's mass. When it reaches 1.44 solar masses, Kaboom! The difference in the Type 1a is that there is no remnant object left. The white dwarf is totally obliterated and, most important for astronomers, the intensity of the explosion is the same for all Type 1a supernovae. This makes it a standard candle which can be used (and is) to determine distances to the farthest reaches of the Universe.

Capricornus – The Sea Goat (July to November)
(Northerly aspect)

Capricornus is not a very bright constellation. That's not to suggest its stars are of low IQ but they are fairly faint. Its brightest star is δ Cap. at mag. 2.9. It represents a goat with a fish tail. I feel it more resembles a large inverted V or a roof cap. The best way to find it (with the help of the monthly star maps) is to look north and it is just south of the mid-way point between Aquila and Aquarius, or east of Sagittarius. The ecliptic passes right through its centre.

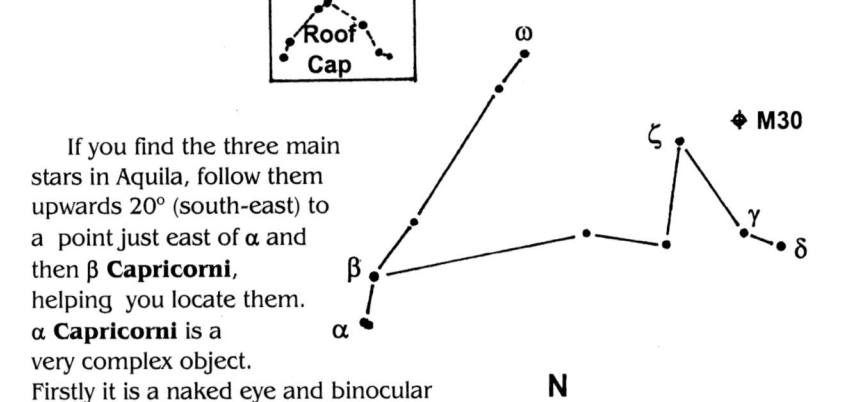

If you find the three main stars in Aquila, follow them upwards 20° (south-east) to a point just east of α and then β **Capricorni**, helping you locate them. α **Capricorni** is a very complex object. Firstly it is a naked eye and binocular optical double, comprising a mag. 4.2 yellow supergiant 690 light years away and a 3.6 magnitude yellow giant only 110 light years away. It is a test of your colour sense to see if you can pick the yellow from the orange star. However, though your binoculars won't show it, each of these stars is another double. The orange star has a wide mag. 9.2 companion while the yellow star has a fainter mag. 11 companion. But there's more. The 11 mag. companion is itself a double. Moderate sized telescopes will reveal all these stars.

Just 2° above (south of) α Cap is β **Capricorni**, another optical double, easily viewable through binoculars. Its yellow-orange naked eye star is magnitude 3.1. In binoculars you will see its magnitude 6.1 blue-white giant companion which, at 850 light years, is almost three times as far away as the other star.

Now find δ Cap at the far right of the 'roof cap', the constellation's brightest star. Just under 2° to its left (west) is γ **Cap** (mag. 3.7). Move 6° upwards (south). That is the location of **M30,** an 8th magnitude globular cluster 30,000 light years away. It is right on the boundary of binocular visibility. It will be a real test for your eyes and binoculars. One reason for this is that it has a highly condensed centre with strings of outward leading stars and not a large object in the sky. Your best chance, obviously, is from a dark site with a clear sky.

Carina - the Keel (December to June)
(Southerly aspect)

This is a treasure trove of binocular objects as it lies in the rich star fields of the Milky Way. Carina represents the keel of the Argo, that great ship sailed by Jason and the Argonauts. In fact, there was originally one super-large constellation called Argo Navis which was the whole ship. However, in 1763 it was split (broken by rocks?) into three more manageable constellations: Carina (the keel), Vela (the sails) and Puppis (the stern). The easiest way to find Carina is to find the False Cross and Canopus. Carina is beneath them. By sweeping this area with your binoculars, you will come across many knots of stars and swirls of gases. You will be amazed.

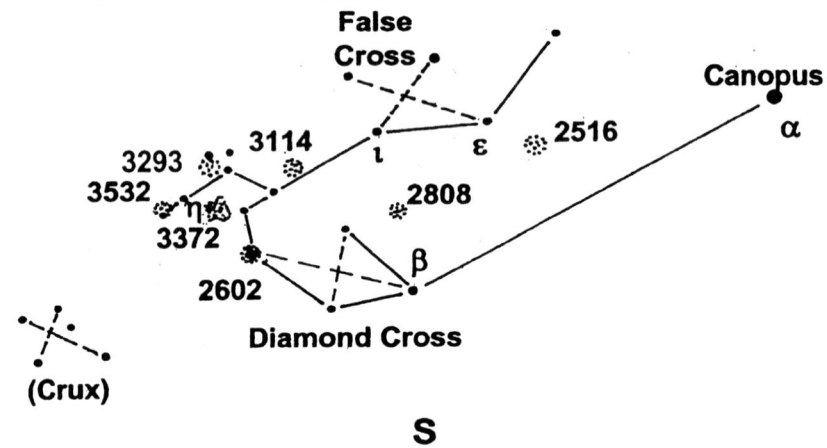

From your monthly star map, find α **Carinae**, or Canopus, the 2nd brightest star in the whole sky at magnitude –0.72 and 313 light years away. Though not any different in binoculars (i.e. it is not a double), it is mentioned here because of its ranking in brightness and it is always useful to recognise it when finding your way amongst the stars. Go east from Canopus and identify the False Cross, a larger version of the Southern Cross. Hanging 3° under the bottom of the False Cross you will see, with your naked eye, a hazy patch. This is **NGC2516**, a delightful cluster of over 80 stars about 1,100 light years away and excellent in binoculars. (It has a nickname – the Southern Beehive.) It covers an area of about two Moon diameters. A giant red star at its centre is visible to small telescopes. To appreciate the size of this cluster which is about 20 light years in diameter, remember that's about 5 times the distance from the Sun to our nearest star, Alpha Centauri.

Now locate the Southern Cross east of the False Cross. You will notice that they are aligned in parallel, a very strange coincidence. Exactly mid-way between them you will find η **(eta) Carinae**, a showpiece object for amateur astronomers. Firstly because it's easy to spot with the naked eye and it looks great in binoculars and small telescopes. Books have been written about η **Carinae**, it is such a fascinating object both historically and scientifically. It is in fact named after the star η **Carinae**, but ironically you can't see the star itself now but the nebula it is buried within. In 1843, it was briefly the second brightest star in the sky, outshining Canopus, but it is now barely visible to the naked eye at magnitude 6. This is because it is a special type of star, a variable nova-like star, which varies in brightness dramatically. The star is believed to be about 100 times as heavy as the Sun and 4 million times brighter. Yet we can't see it because of the huge cloud (or nebula) of dust and debris thrown off by its flare up back in 1843.

What we do see is a most beautiful nebula (which has the title **NGC3372**), one of the most popular photo-opportunities for amateurs. It appears as a large whitish cloud of gas and faint stars, broken into large triangular segments by dark lanes. That cloud is being lit by the powerful, but to us invisible, star within. The star and nebula are estimated to be 7,500 light years away. You will find that you'll keep coming back to this one.

All around η **Carinae** is a gaggle of beautiful knots and clusters. Here are some of the more popular ones.

About 2° to the N.W. of η **Carinae** is nice tight knot of stars called **NGC 3293**. Then 4.5° to the west of η **Carinae** is **NGC3114**, also visible to the naked eye, but being 3,000 light years away, over twice the distance as NGC2516, its stars are a lot fainter. It is pretty, but not as distinctive as 2516.

Then find η **Carinae** again and move 2.5° to its east. There you will see **NGC3532**, a superb open cluster, containing over 150 stars, about 1,400 light years away. Looking like a fuzzy magnitude 3 star to your naked eye amongst the Milky Way's rich star field, binoculars reveal a delightful patch of stars covering an area about 1° x ½°. The stars seem to form patterns and sets of curved and straight lines.

Finally (or is it), 4° below η **Carinae** is **IC2602**, another open cluster, also about 1° in diameter. At 480 light years away, you can see its brighter stars with the naked eye, while to the binoculars there are about 60 stars. It is said to resemble the famous Pleiades, only a lot smaller, and is often called the Southern Pleiades. You be the judge. Another feature about **IC2602** is that it is the top component of yet another 'cross'. Hanging below **IC2602** are three stars in a cross shape parallel to the Southern and False Crosses. I have heard this referred to as the Diamond Cross. It is one of those strange coincidences that around Easter time, there are three upright crosses hanging overhead, like the three crosses at Calvary.

If you run a line between the eastern star of the False Cross (ι) and the southern star (β) of the 'Diamond' Cross, about midway you will find a 6th magnitude globular cluster, **NGC2808**, very easy in binoculars as a fuzzy spot. Obviously it is a long way away.

That's not all though. The area around η **Carinae** and moving towards the False Cross is rich with stars. Take the time to enjoy the various groupings.

Centaurus – The Centaur (February to September)
(A southerly aspect)

This is another happy hunting ground for binoculars, as well as containing a large number of astronomically significant objects. Centaurus is a huge constellation, covering an area of about 60° x 40°. The sketch below gives an idea of where the shape of the centaur is, but there are so many stars in it and it is so large, it's not easy to identify at a casual glance. You have to follow the sketch and map it out. (Not a bad idea, by the way.) As a general hint, Centaurus straddles the Southern Cross and towers above and to the left of it.

The two Pointers are in fact the front hooves of the centaur, with one rear hoof on the opposite side of the Cross.

The constellation's name springs from Greek mythology. It represents the famous half man-half horse, made popular in the Fantasia movie.

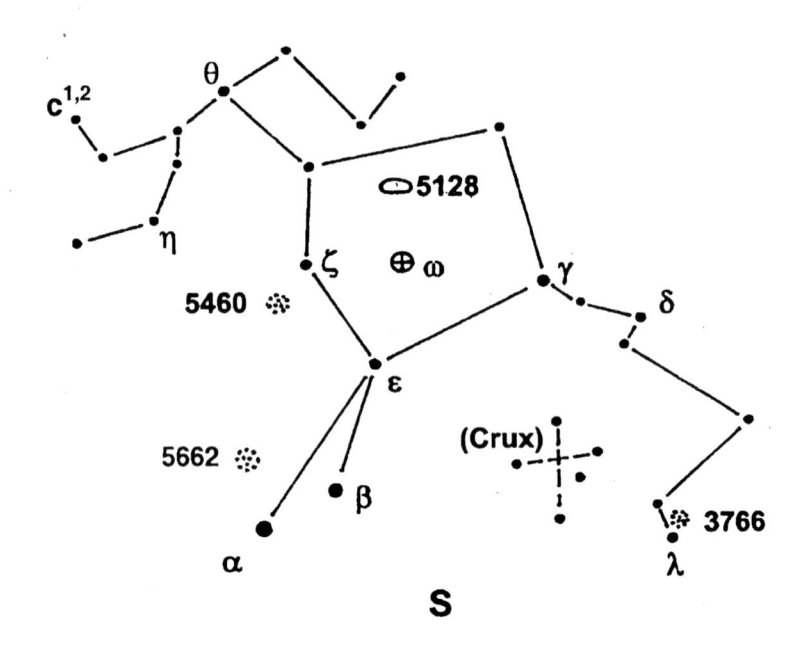

α **Centauri** is the Pointer star furthest from the Southern Cross. It has the name Rigil Kentaurus, meaning 'the foot of the centaur'. It is not a binocular object (even though it is a magnificent binary) but is included here because of its immense interest and prime importance to us. Not only is it the third brightest star in the sky at mag. -0.27, but it is the closest star to our Sun, at a distance of 4.3 light years. Our nearest neighbour. G'day Rigil, shake my hoof.

α **Centauri** is in fact a triple star system. The two main components are yellow stars similar to our Sun, but they orbit very close to each other – about the same distance that Saturn orbits our Sun – and do so every 80 years. As a result, even though they are 'close' to us, binoculars still don't have sufficient magnification to split them apart. But even the smallest of telescopes will show α **Centauri** as a bright pair of stars, very close together. They look like car headlights coming at you from a long way away.

The third star in the system is called **Proxima Centauri**. It orbits the other two stars at a distance of about 1/6th of a light year and takes 1 million years for each orbit. Proxima is a very faint 11th magnitude red dwarf star and not visible in binoculars. Its only interest to us is that it is at present 0.1 light year closer to us than the other two components of α Centauri so, technically, it is the closest star to our Sun.

ω **(Omega) Centauri** is a major **Wow!** object. It is the biggest globular cluster in the sky. It looks magnificent in binoculars, like a ball of fairy floss, or a street light with a thousand bogong moths fluttering around it. Even in a poor night sky, you can still see this superb island 'universe' of ancient stars.

It is very easy to find. In a dark sky it actually can be seen as a fuzzy 4th magnitude star. First go from the 2nd Pointer star, Beta Centauri, roughly parallel to the line of the Southern Cross to the next bright star, **Epsilon (ε) Centauri**. Keep following that line through ε for about ¾ the distance and there you will find ω. It is very hard to miss.

What makes ω **Centauri** all the more fascinating is knowing that it is 17,000 light years away. Believe it or not, that makes it one of the nearest globular clusters. I guarantee once you have found this, it is so easy and awesome you'll keep coming back to it.

Then, continue the line from ε through ω for another 5° (roughly ¾ the distance from ε to ω), then east about 1°. That's the location of **NGC5128**, also known as **Centaurus A**, an active radio galaxy 13 million light years away. In a clear dark sky, you have a very good chance of seeing it in your binoculars. What you will see will be a faint hazy circular patch with a large dark divide diagonally down the middle, a far cry from the beautifully detailed photos we see in books. You will probably need to use averted vision to convince yourself you are actually seeing it. But as faint and indistinct as it may be, remember… 13 million light years.

Now go back to ω **Centauri**. Move eastward 7° until you are directly above (north) of β **Centauri**. There you should find **NGC5460**, an open cluster with about 40 stars covering an area as large as the Moon. On the limit of naked eye visibility, it shows up nicely in binoculars.

While at the α - β end of Centaurus, there is another open cluster, **NGC 5662**, which forms a triangle with those two stars, though it is slightly closer to α than β. There are about 50 stars in the cluster and they form a sort of triangular patch with a suggestion of a gap down the middle.

Finally, let's find another object, **NGC3766**. It is an open cluster of about 100 stars 6,300 light years away and is located 1.5° north of the star λ **Centauri**, the rear hoof west of α Crux, and can be seen as a fuzzy patch with the naked eye. You will know you have found this cluster in your binoculars when you see the short line (1° long) of three stars in the same field of view, 'pointing' to it.

If you follow the line from α **Crux** to δ **Crux** by twice that length, you find a delightful triple star arrangement δ **Centauri**. While only an optical multiple, this is a tidy triangle (with sides of about 4') of all blue-white stars of magnitudes varying from 2.6 to 6.4 and distances from 99 to 522 light years. The brightest star of the trio, δ, is itself an Algol type eclipsing binary, but its magnitude variation is too small for amateurs to detect.

If you now locate η and θ **Centauri** and imagine an equilateral triangle with them as a side, the triangle's other point will be at a wide optical double c^1 and c^2 **Centauri**. This attractive pair comprises a mag. 4 red star 230 light years away and a mag. 5 blue-white only 55 light years away.

How long do stars live?

From the moment when it is first formed, a star's temperature and lifetime is determined by its mass. Simply put, the more massive a star is, the shorter its life. The less massive, the longer its life. This may seem counter-intuitive. You'd think "the more 'stuff' there is, the longer it takes to use it up". But in fact, the more 'stuff' there is, the greater the gravitational pressure at the core, the hotter the core, and the more furious the rate of fusion, using up its hydrogen faster. Our Sun – an 'average sized' star, at one solar mass (by definition) – should live for around 10 billion years. A much larger star, say 10 times the Sun's mass, would be very hot and blue-white in colour, and last about 100 million years. The biggest stars (there is a theoretical maximum of 100 solar masses) would be superhot blues and last less than 10 million years.

At the other end of the scale, stars smaller than our Sun, say $1/10^{th}$ its mass, would never get above the red colour but would have incredibly long lives, even thousands of billions of years – probably as long as the Universe.

Cetus – the Whale (September to January)
(Northerly aspect, but high.)

This is a large constellation made up of relatively faint stars. Its brightest star (β, not α) is mag. 2.0. It represents a sea monster (surely not a lovable whale?) from Greek mythology which was ravaging the seaboard and about to eat Princess Andromeda when Perseus slew it using the head of Medusa. For binoculars, it is bereft of spectacular deep space objects, but it does have one fine double and two 'ordinary' stars of very special interest.

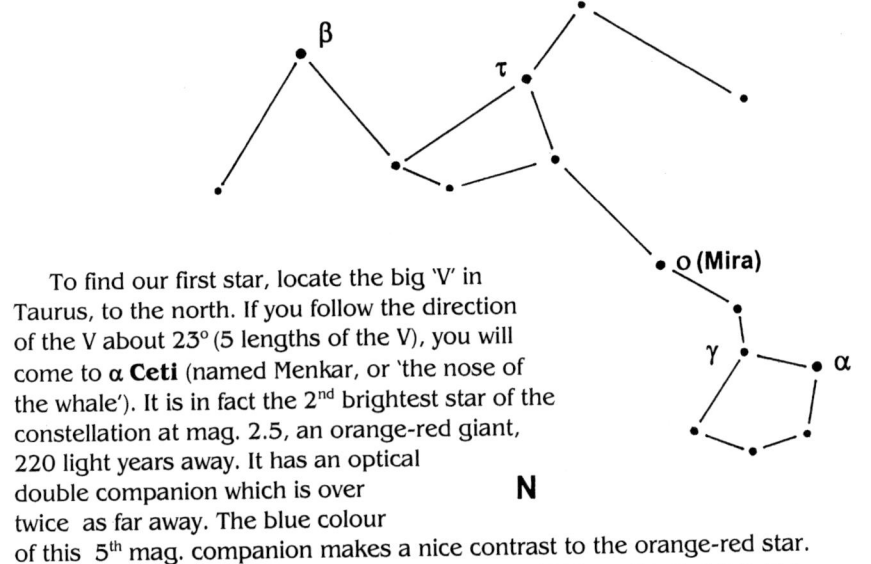

To find our first star, locate the big 'V' in Taurus, to the north. If you follow the direction of the V about 23° (5 lengths of the V), you will come to α **Ceti** (named Menkar, or 'the nose of the whale'). It is in fact the 2nd brightest star of the constellation at mag. 2.5, an orange-red giant, 220 light years away. It has an optical double companion which is over twice as far away. The blue colour of this 5th mag. companion makes a nice contrast to the orange-red star.

At the opposite end of the whale (33° west and 23° south), find β **Ceti** (mag. 2). 15° to the east and down (north) 2° from β is τ **(Tau) Ceti.** It is very interesting, though a very ordinary mag. 3.5 star. Its interest lay in the fact that it is almost identical to our Sun, is a single star (like our Sun) and is only 11.9 light years away. That makes it the closest Sun-like star to us and a prime candidate for a life-bearing solar system. Watch that space!

Halfway down the neck of the sea monster, you will see a star, its magnitude depending on when you are looking. This is the famous **Omicron Ceti**, or **Mira**, the 'Amazing (or Wonderful) One'. It is amazing because it was the very first discovered long period 'variable star'. Mira varies from 3rd to 9th magnitude over a period of 332 days. So if you observe this star when it first appears over your horizon and continue to watch it over the next 6 months (before it disappears until the next year), you should see a dramatic change in its brightness, from visible to your naked eye to barely visible in your binoculars. Amazing!

Chamaeleon – the Chameleon
(Southerly aspect)   (All year, but best January to July)

An uninspiring constellation of faint stars named by Dutch navigators in the late 16th century. Their nights at sea must have been remarkably boring. It is very close to the South Celestial Pole, immediately beneath Carina, the Keel. It looks like a long stretched kite, on its side. To find it, extend the length of the Southern Cross 2.5 times. This should land very close to β **Cham**.

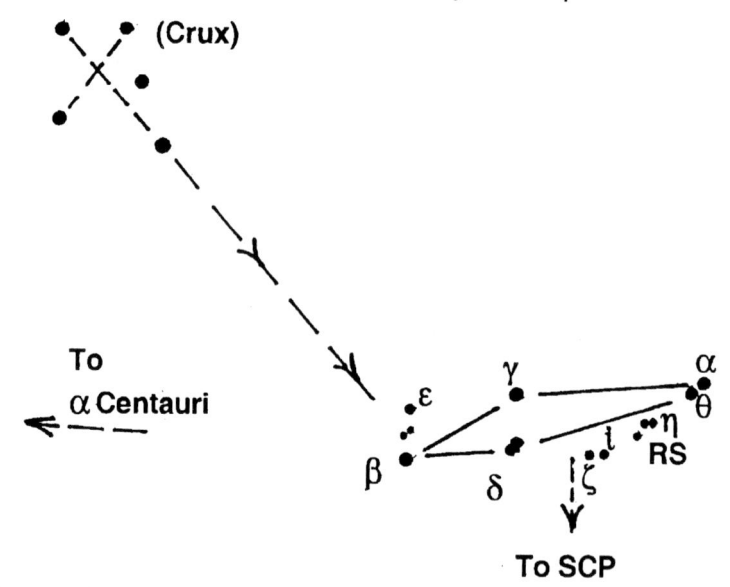

It has a variety of binocular optical doubles of interest. Firstly the attractive δ **(delta) Cham,** the bottom star of the kite. Its two unrelated stars, 4.5' apart, show up clearly in binoculars. One is a 5th mag. orange giant 350 light years away while the other is a 4th mag. blue star 10 light years further away. Again, note the contrast in colours.

The western end of the 'kite' comprises two separate stars that make an attractive wide double in binoculars. α (alpha) and θ (theta) Cham. are white (mag. 4.1) and orange (mag. 4.3) and are actually quite close to each other at 63 and 78 light years respectively. Then spread out between alpha and delta are two other optical double pairs: ζ (zeta) and ι (iota) Cham. are both 5th magnitude, white (718 light years) and yellow (294 light years). Then η (eta) and **RS** Cham are both 5th magnitude.

Finally (?) at the eastern end, just 1.5° above β **Cham.** is a beautiful blue and gold wide double. It's ε **Cham.** Though fainter (mags. 4.9 and 6.5) and closer (only 2.3' apart) than δ, your binoculars will split them into a very delicate and attractive pair. And look for the even fainter (mag. 7) and wider pair between β and ε.

Coma Berenices – Berenice's Hair (April to June)
(Northerly aspect)

This is another faint constellation. It contains many galaxies (the Coma Galaxy Cluster) that can be seen through amateur telescopes but, unfortunately, not binoculars. However, to compensate, it contains a delightful cluster of stars that can only be appreciated through binoculars, as they cover a large area of the sky. The cluster is a delightfully delicate sprinkling of stars. The constellation represents the tresses of Queen Berenice's much admired hair after she cut them off in thanksgiving for the safe return of her husband, the Pharaoh king.

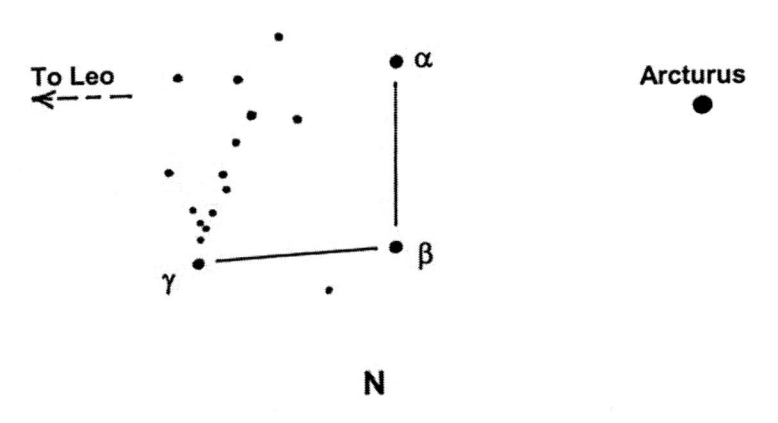

The three brightest stars of the constellation, α, β, and γ, are faint at mags. 4.3 and 4.4. These form a very faint right angle triangle. First find that. It lies generally 15° west of Arcturus in Bootes.

The even fainter 5th mag. stars that comprise the 'tresses' are sprinkled above (that is, to the south) of the western star, γ. To the naked eye there is an impression of a 'faint cloud', very obvious from a dark site. Overall, there are about 50 stars in the group, but many will be too faint for binoculars. However, you should be able to pick out the tresses' main stars that have a distinctive 'V' shape. The cluster's centre is about 280 light years away. For binoculars with a 5° field of view, the Coma Berenice's Star Cluster should occupy your whole field. Enjoy!

Corona Australis – the Southern Crown

(Southerly aspect, high overhead.) (June to October)

There are two 'crowns' in the sky and we can see both, but the Southern one is high overhead and easier to see. Though made of faint 4th mag. stars, you can easily spot the distinctive circlet immediately below (south of) the base of the Teapot in Sagittarius. Various mythological legends have it as a crown or a circlet like the Caesars wore.

This crown is about 3° tall so should fit into your full field of view.

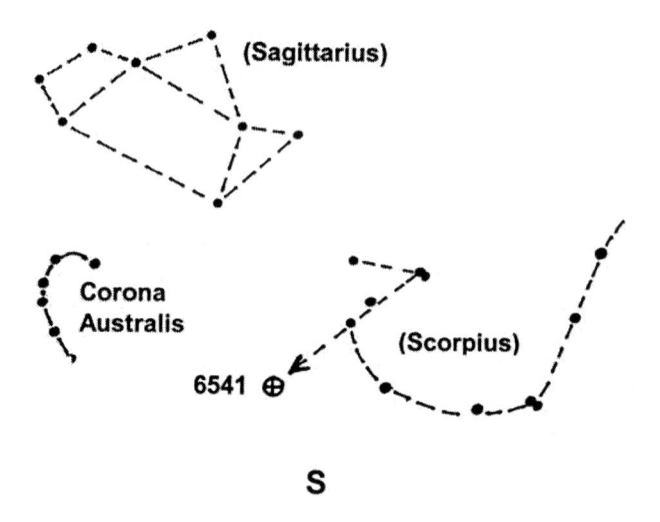

NGC6541 is a 6th mag. globular cluster. It is 22,000 light years away making it one of the closer ones. An easy way to find it is to use the hooked tail of Scorpius as a pointer. Follow the last three stars of the tail backwards the same distance as the tail and it should be in the centre of your binocular field of view.

From a moderately light polluted backyard, I can see it as a small fuzzy patch with a concentrated centre, beside another 6th magnitude star. A darker sky would show it up better.

Though it appears closer to Scorpius, **NGC6541** is technically in the constellation Corona Australis.

There's a pretty little sight inside the circlet. From the left side, cutting across the middle of the circlet left to right, there is a smooth curve (like a sine wave) of at least ten 6th and 7th magnitude stars. It gives me the impression of a large open mouth (the circlet) with a long wagging tongue protruding. See what you think of it.

Corona Borealis – the Northern Crown

(Low northerly aspect) (May to August)

This is the other crown. It has a more regular and distinct shape. Because of its proximity to the northern horizon, you'll need a clear non light-polluted horizon. Apart from its bright 2nd mag. α star, the other six stars in the crown are fainter at 4th magnitude. It is easiest found by sweeping about 20° east of and 10° below (north of) Arcturus in Bootes in the direction of Hercules.

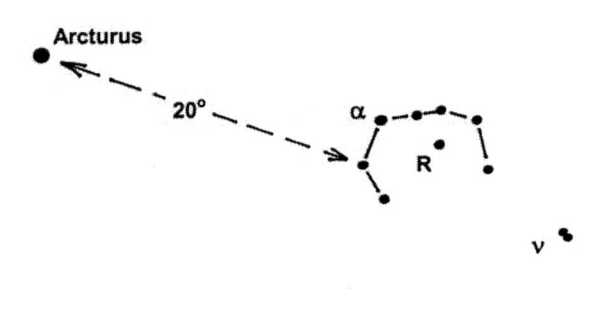

N

Unfortunately, from a binocular perspective, Corona Borealis is famous for something that binoculars cannot see. That is, a giant cluster of some very distant galaxies, about one billion light years away. As you gaze at this faint semi-circle of stars, remember that invisible in your field of view are over 400 galaxies about 1/10th of a universe away. **Wow!**

On the eastern fringe of the circlet and about 5° lower is ν **(nu) CrB**, a wide optical double star comprising two 5th mag. orange giants. They should make an attractive pair of orange cat eyes in your binoculars.

One star to keep an eye on is a 6th mag. star **R CrB** found almost in the centre of the crown. This is an irregular variable, a yellow supergiant that will, without warning, start to fade over a period of a month till it is invisible in binoculars (down to mag. 14). Then, just as quickly, it will reappear and gradually return to mag. 6. You never know when this is going to occur, so it pays whenever Corona Borealis is in view, to have a peek and see what's happening. Another opportunity to be a Variable Star watcher.

Crux – the Southern Cross

(Southerly aspect)   (All year, but best March to August)

This is the constellation featured on the Australian and New Zealand flags. No instructions on finding it are necessary, except... don't confuse it with the larger False Cross to its west. As a general rule, face south, locate the two bright Pointer Stars and follow them to the top of the Cross.

Because of its relatively small size (6° x 4°), a pair of binoculars with a field of view of 7° (eg 7x50s) should fit the whole Cross in.

While, scanning the Cross, if it's a good dark sky, check out the Coal Sack, a famous large dark nebula that nestles to the left of α and β Crucis. Like a shadow puppet, it is visible only because of the star light it is blocking out from behind. The visibility of the Coal Sack is usually a sign of a good dark sky.

α **Crucis**, the bottom star, at mag. 0.87, is the 14th brightest star in the sky. It is a true binary, with bright blue-white component stars of mags. 1.3 and 1.7. Though over 320 light years away, they are too close to be split by binoculars. However, binoculars do reveal two stars, one much fainter than the other. The fainter star is an unrelated 5th mag. optical companion of the binary pair.

γ **Crucis**, the top star, is the odd star out of the Cross, a mag. 1.6 giant red star compared to the other blue-whites. It forms a wide optical double with a 6th magnitude white star. They are easily split in binoculars.

(NOTE: A question commonly asked is if the four stars of the Cross are the same distance away. The answer is resoundingly 'No'. α, β, γ and δ are, respectively, 320, 335, 88 and 364 light years way. Note that γ is the odd one *really* out.)

NGC4755 is the pride of the Southern Cross and a favourite of most (southern) amateur astronomers. Located 0.5° south-east of β Crucis and appearing to the naked eye as a faint fuzzy star κ **Crucis**, this famous open cluster bears the name 'Jewel Box', and deservedly so. Though appearing small and compact in binoculars (it is 5,000 light years away), with a distinct arrowhead or capital 'A' shape, it is arguably one of the most beautiful star clusters in the sky, north or south. It contains over 50 stars, mostly blue-white giants (like sapphires) but there are yellows (topaz) and one bright red supergiant (a ruby) at its center (on the cross bar of the capital 'A').

Once you've seen it in your binoculars, find a friend with a telescope and have a closer look. It is absolutely beautiful. This is another object you'll keep coming back to.

Cygnus – the Swan (August to October)
(Low northerly aspect)

This constellation is quite large, with the southern parts easily observable but the more northern objects very close to our horizon. For binoculars, it has a grab bag of double stars, a cluster and an interesting nebula.

On top of that, despite its representation of a swan in full flight (in mythology, it was a guise Zeus used to visit the beautiful Leda on her wedding night to another man), it is also known (mostly to those in the other hemisphere) as the Northern Cross. This cross is made up of the stars α (Deneb), β (Albireo), δ and ε Cygni. It is not so obvious to see the cross, as it is much larger than the Southern (real) Cross.

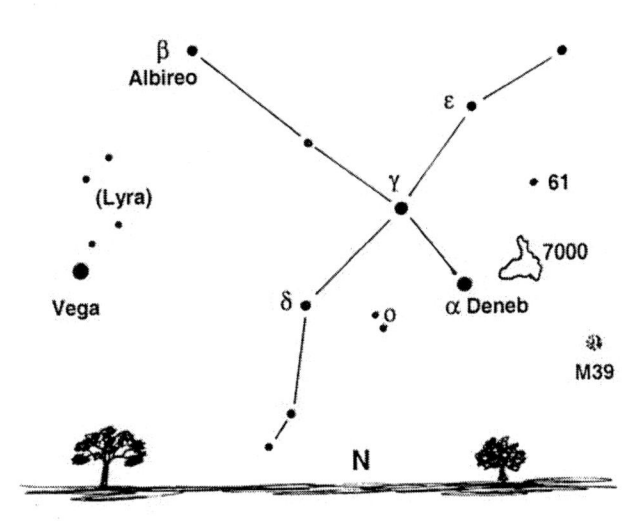

α **Cygni** (Deneb), the brightest star in Cygnus at mag. 1.3 and the 19th brightest in the sky, while not a binocular interest object, is worth gazing at while knowing that it is a blue-white supergiant, over 25 times more massive than our Sun and 60,000 times more luminous. When this star goes supernova, and one millenium it will, a black hole will be left behind. **Wow!**

The next objects are in the vicinity of α Cygni:

From α, move 2° down (north) and 4° west to find o **(omicron) Cygni**, a beautiful optical double, ideal for binoculars. You will easily see the mag. 3.8 orange and 4.9 turquoise components. It's like Albireo (see later), only the wider separation is easier to see and it's not a true binary pair. If you can hold your higher power (say 10x) binoculars very steady, you may be able to spot the 7th mag. blue star that accompanies the orange star (a binary pair). Omicron, like Deneb, is very near the horizon so it will be tricky to see from very southern latitudes.

Move back to α Cygni, then about 3° directly east to see **NGC7000**, called the North America Nebula, for obvious reasons when you see the time exposure photographs. The dark Mexican Gulf stands out starkly against the glowing nebulosity. In binoculars, the nebula doesn't stand out that well but you can still detect it with the naked eye or binoculars as an area of increased brightness in that part of the Milky Way. Covering an area of about three Moon diameters, binoculars will show a glowing patch suggesting a North America continent shape if the sky is dark and clear enough. But remember, to us it will look upside-down.

If you move 6° directly above (south of) NGC7000, you'll find a 5th mag. orange star, **61 Cygni**. 61 Cygni is interesting on two counts. Firstly it is a binary comprising two orange dwarf stars each about half the size of the Sun. They orbit each other every 650 years. Being only 11.4 light years away (the 15th closest star), the two stars can be split by large magnification binoculars if held steadily.

Secondly, 61 Cygni holds a place in astronomy history as being the first star to have its distance accurately calculated directly using the painstaking parallax method. This was done in 1838.

Go back to α Cygni, drop down (north) by 3°, then right (east) by 9°. You will find **M39,** a fairly loose open cluster of about 30 stars of 7th mag. under 1,000 light years away. They cover a Moon sized area of sky. This is also very low on the northern horizon seen from southern latitudes.

Finally, all by itself at the beak end of the swan, is a most beautiful star. To find it, locate the 0 mag. star Vega (in Lyra) from the monthly star maps. To the right and above Vega is a tall parallelogram. Follow the line between the top two stars east by 3.5 times its length. You will there find β **Cygni** (Albireo), a beautiful mag. 3.1 binary star 300 light years away. In a small telescope, it appears as a glorious colourful gold and blue-green pair, like party lights in the sky. A high magnification pair of binoculars (say 16x), if held rock steady (see Section 5 for tips on how to do this) should just split them. Make sure you are perfectly focused in both eye pieces as this is a difficult double to split. Average magnification binoculars will, unfortunately, not split Albireo, so take any opportunity to view it through a friend's telescope – it is beautiful.

Delphinus – the Dolphin. (July to November)
(Northerly aspect)

This very small constellation represents that noble creature, the dolphin, based on their mythological role as couriers for Poseidon, the sea god. The main feature of this constellation, easily recognised and found east of and just below Altair (in Aquila), is the 'asterism' (i.e. a group of stars) that looks like an elongated parallelogram, or a kite.

There are no special binocular objects in Delphinus apart from some wide optical double stars, but the 'kite' is very attractive and being just 3° long, fits neatly into the field of view. It also has the unusual name "Job's Coffin", source unknown.

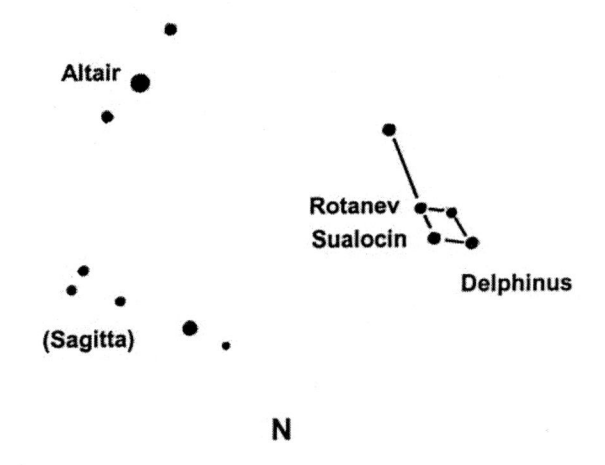

```
                    •
       Altair  ●
           •
                              •
                              \
                  Rotanev  ●—●
                  Sualocin ●—●
                                   Delphinus
             •
            •  •
       (Sagitta)    •
                       •

                    N
```

As a piece of interesting trivia, the names of the two main stars contain a cheeky joke. The top star is Rotanev while the left hand one is called Sualocin. Very unusual names, you'd agree. Now turn the letters around and you have Nicolaus Venator. You scholars will of course recognise this as the latinised name of Niccolo Cacciatore. This gentleman was the brilliant assistant to the Astronomer at Palermo Observatory in the 1800s and obviously managed to get his name immortalised in the sky by inserting them in the Palermo catalogue in 1814. Good luck to him. His cheek was later rewarded by being appointed Astronomer at Palermo.

When you are enjoying the full view of Delphinus in your binoculars, note the wide doubles that accompany both Rotanev, Saulocin and also the bottom right star, γ Delphini. Rotanev's and Saulocin's companions are both white stars, magnitude 5 and 6 respectively. Gamma's is a fainter red star at 7th magnitude.

Dorado – the Goldfish

(Southerly aspect)   (All year but best from November to May)

This is a very ordinary constellation that is dominated by its main resident, the only object of binocular interest. But what an object!

Large Magellanic Cloud (LMC) is easily spotted by the naked eye in a dark sky as that 'cloud' about 6° in diameter that isn't moving. It looks like a piece of the Milky Way that has drifted off. (Warning: There is another of these, the Small Magellanic Cloud covered in Tucana. The LMC is the larger of the two.) It is found approximately halfway between the South Celestial Pole and the star Canopus.

LMC is one of the two nearest galaxies external to ours. In a sense, it is a mini-galaxy, orbiting our own. There is a current theory that it is slowly being broken up by the Milky Way's gravity and it will eventually be absorbed into the Milky Way.

LMC contains about 10 billion stars and is of an irregular shape. That is, it's neither a spiral nor an ellipse. LMC is 170,000 light years away and about 30,000 light years in diameter. Try to imagine its size and proximity to our Milky Way, remembering that the Milky way is 100,000 light years in diameter. Study of this large cloud reveals individual stars and numerous nebulae. The darker your sky the better.

When studying LMC in your binoculars, what will become immediately obvious (you can't miss it) is the presence of a very bright appendage at the bottom left of the LMC – **the Tarantula Nebula**. Tarantula Nebula, also called Dorado 30, is a delight to observe in binoculars. It is part of the LMC and so is also 170,000 light years away. To binoculars it becomes a large bright area of glowing gas. Within that nebula, giant stars are being born and more than 100 existing newly born supergiant stars are belching out the light that causes the nebula to glow.

Tarantula Nebula (it gets its name from the dark lanes that appear as hairy spider legs in photographs), is 1,000 light years in diameter and is one of, if not the, largest nebulae of its type in the known universe. If it was located in our Milky Way at the same distance as, say, Orion Nebula, it would be three times brighter than Venus and cover an area of 30°, about the same size as the constellation Scorpius. ***Wow!***

Equuleus – The Little Horse (July to November)
(Northerly aspect)

Not so much a little horse, as a small part of one – its head. Not an inspiring constellation, containing only faint stars and no deep space objects. (In fact, it is only included here for completeness.) It is easiest found by following a line from Sagitta through Delphinus for an equal distance. It is also immediately west of its 'big brother' Pegasus, though there is no known mythological link between the two. After Crux, it is the second smallest constellation in the sky.

γ **Equulei** is an optical double star, consisting of a white star (mag. 4.7, 115 light years away) and an unrelated white mag. 6 star called 6 Equulei.

If you scan the length of the constellation, you will note a number of other faint optical doubles.

Canopus, in Carina, is a white supergiant star many times more massive than our Sun. This is one of the few stars named after a real person. When Helen eloped with Paris, starting the famous Trojan War, her husband the Spartan King Menelaus launched a huge fleet against Troy to recover his wife. The pilot of that fleet was Canopus. After the war, Canopus performed am epic feat of navigation to return the fleet safely to Egypt. Unfortunately, on stepping ashore, Canopus was bitten by a snake and died. King Menelaus named the star in Canopus' honour.

It is fitting that Canopus is a key star used by astronauts for their navigation in space.

Gemini – the Twins (January to April)
(Northerly aspect)

The twins in question, Castor and Pollux of Greek Argonaut mythology, are its two main stars which stand out quite brightly at mag. 1.6 and 1.2, and when up, are easy to recognise. It is also useful to know that they are **exactly** 4.5° apart so you can use this to judge other angular distances in the sky. For example, 9° is exactly twice the distance between Castor and Pollux.

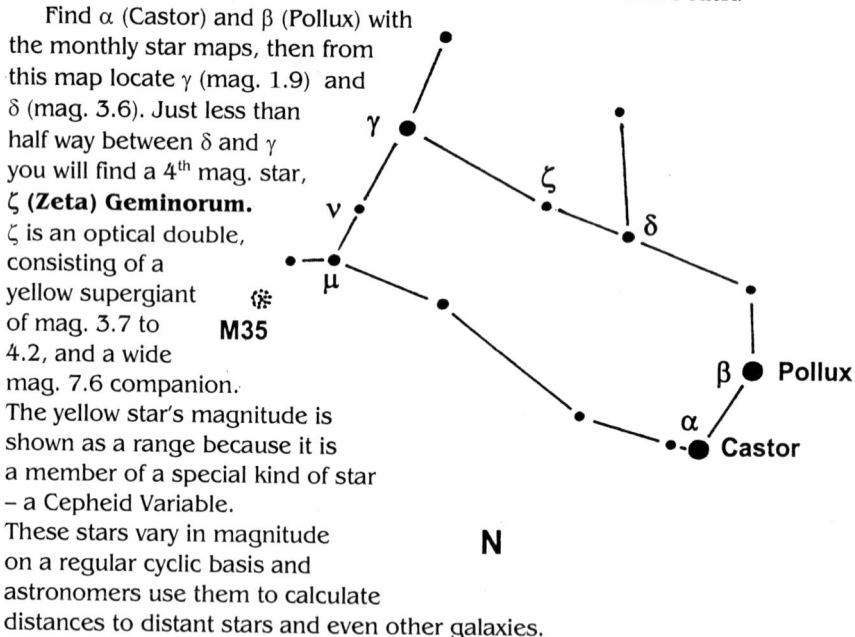

Find α (Castor) and β (Pollux) with the monthly star maps, then from this map locate γ (mag. 1.9) and δ (mag. 3.6). Just less than half way between δ and γ you will find a 4th mag. star, ζ **(Zeta) Geminorum.** ζ is an optical double, consisting of a yellow supergiant of mag. 3.7 to 4.2, and a wide mag. 7.6 companion. The yellow star's magnitude is shown as a range because it is a member of a special kind of star – a Cepheid Variable. These stars vary in magnitude on a regular cyclic basis and astronomers use them to calculate distances to distant stars and even other galaxies.

In Zeta Gem's case, it varies over 10.2 days. As it varies by 0.5 magnitude, you may be able to pick it up by watching its brightness compared to δ **Gem** (mag. 3.6) over ten days. If you can see a difference, congratulations… you have become a Variable Star Watcher, an important breed of amateur astronomers.

Now find the 3rd mag. star μ **Gem** (from map), about 7° below and to the west of γ. Just over 1/3rd the way from μ to γ you will find the 4th mag. star ν **(nu) Gem,** another optical double, consisting of a hot blue-white star 500 light years away, with a wide but very faint mag. 8.7 companion just visible in binoculars.

Starting from μ **Gem** again, move 3° north-west to find **M35**, a lovely open star cluster containing about 200 stars. At 5th magnitude, it is visible to the naked eye as a fuzzy patch about as big as the Moon. Binoculars will resolve it into stars and you should be able to see a number of stars linking like looping chains. Look at it long enough and you will start to see all kinds of shapes and patterns. This cluster is 2,800 light years away.

Grus – the Crane (July to January)

(Southerly aspect, overhead)

There is a certain similarity between the shape of Grus and Cygnus – after all, they are both long necked birds. The main difference is that Cygnus covers about twice the area of sky. For binoculars, Grus contains three nice optical doubles, all above (north of) the two brightest stars α and β Gruis.

To find Grus, first find the big Y in Aquarius. 30° due south of that you will find a bright 1st magnitude star, Fomalhaut. (Fomalhaut is in Piscis Austrinus which is not covered in this book.) Continue 15° south to find β Gruis in the centre of Grus. (It will be on your monthly star map.) For a shorter cut, face south and, depending on the month or time, it will be almost directly overhead. Grus actually does look like a long necked bird in flight, but with its neck bent towards the west.

1° above the mid-point between α and β **Gruis** is π **(pi) Gruis**, an optical double, divisible in your binoculars, though faint. The red star is a variable (but not a Cepheid) changing from mag. 5.4 to 6.7 over about 150 days. This means that it is sometimes fainter or brighter than its line of sight companion, a giant white star of mag. 5.6.

Then, 2.5° above and 1° left (east) of π is δ **Gruis** (in the neck), an optical double that can be seen by the naked eye. Your binoculars separate the mag. 4 yellow giant and mag. 4.1 red giant even further. They are 295 and 325 light years apart respectively.

Finally, from δ **Gruis**, move 2.5° up (north) and right (west) up the neck to locate μ **Gruis**, also a naked eye optical double. Both stars are 5th magnitude yellow giants, 260 and 240 light years away.

As you scan the crane's neck (and you have to crane your own neck to do it), you will spot off to the sides a number of other fainter doubles, and a faint triangular trio just short of γ at the beak.

Hercules

(June to August)

(Low northerly aspect)

Hercules is a large constellation, representing a kneeling man. Different cultures claim this man as one of theirs but it is generally accepted as he of the famous 'Twelve Labours'. Broadly, it can be found between Lyra in the east, and Corona Borealis in the west, and immediately below (i.e. to the north) of Ophiuchus.

What is helpful is that this constellation still appears 'right way up' to us in the southern hemisphere, probably because it is mainly directly overhead for northern people and they looked at it from the same direction that we do over the equator. The trapezoid which is Hercules' pelvis is commonly known as 'the Keystone'.

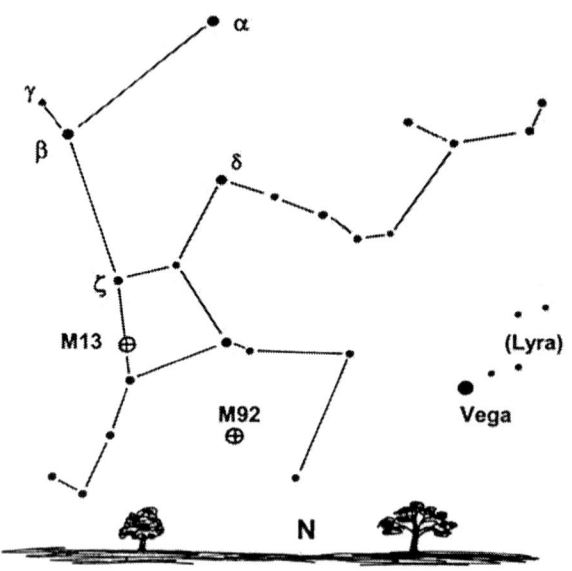

α **Herculis** is not a binocular object but still a very interesting star. It represents the head of the kneeler (i.e. Hercules) and it is one of the largest stars known. At a distance of 400 light years, it is a red supergiant with magnitude varying between 3 and 4, and is over 400 times the Sun's diameter. That is, it has a diameter over 560 million km. If it was placed at the centre of our solar system, it would consume the asteroid belt between Mars and Jupiter. ***Wow!***

Move down and to the left (west) of α to find β **Her.**, then down (north) by 10° to find ζ **(zeta) Her**. Both β and ζ are mag. 2.8, the brightest stars in Hercules. Move another 4° below ζ (along the western side of the Keystone) and you will see **M13**, the globular cluster the Northern hemisphere claims as their greatest. Barely visible (if at all) to the naked eye at 6th magnitude, it is quite striking in binoculars as a misty glow covering an area of about one quarter of our Moon. This misty glow, however, contains over 300,000 stars, is over 100 light years in diameter and is 25,000 light years away. When you observe **M13**, you will note the difference in appearance to other globular clusters. **M13** does not have such a concentrated centre as many others do (e.g. 47 Tucanae). Though classed as a Messier object, **M13** was in fact first identified by Halley (of that comet fame) in 1714.

Now move 7° east of M13, then 6° down (north) below the Keystone to find **M92**, another beautiful globular cluster easily found in binoculars. At mag. 6.5, it cannot be spotted with the naked eye. It is closer to the northern horizon than M13 and will look like a fuzzy object but with a more concentrated centre than M13. M92 is 29,000 light years away.

What is the Zodiac?

Firstly, astronomy should never be confused with astrology, which has no scientific basis. In astrology, the Zodiac is that belt of twelve constellations that lie in the path the Sun takes across the sky. This path is in the plane of the Earth's orbit about the Sun and is called the Ecliptic.

Starting from a point in the sky called the Vernal Equinox (which is where the Ecliptic intersects the Celestial Equator in the Northern Hemisphere), the 18° wide band of the Zodiac is divided equally into twelve sections 30° long. Each section is meant to contain one constellation, or Sign, of the Zodiac, but due to the different sizes of these constellations, some take more or less than their allotted 30°.

The Zodiacal constellations are, starting from the First Point: Aries, Taurus, Gemini, Cancer, Leo, Virgo, Libra, Scorpius (not Scorpio), Sagittarius, Capricornus, Aquarius, and Pisces. Your 'star sign' comes from that period when the Sun is in that part of the sign. The word Zodiac comes from the Greek word for 'zoo' as most of its constellations have to do with animals.

Due to the Earth's precession (wobble) about its axis, the Vernal Equinox has moved from Aries and is currently in Pisces. Over 26,000 years, it will have moved through all the zodiac and back to Aries.

Hydra – the Water Snake (January to May)
(Overhead and to North)

Considering this is the largest (actually the longest) constellation, it is surprising that there is really only one worthy binocular object in it.

Representing the mythical water monster slain by Hercules, Hydra spans a bit more than half the sky, being 100° long. It comprises mostly faint stars so is difficult to trace out. The only significant pattern in it is the head, made up of six stars in a rough pentagon shape the size of the Southern Cross. For that reason, the map below will only show the head end of the constellation. You can try to trace out the complete constellation using your monthly star map. It slithers south of Crater and Corvus and ends just short of Libra.

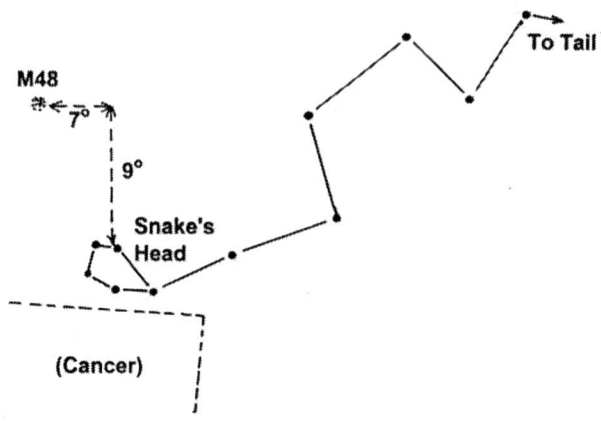

Find the snake's head while facing north. To do this, find Cancer and the head will be directly above it (south). Move up 9° and left about 7° to find **M48**, an attractive open cluster of about 80 stars. On a dark clear night, it should be just visible by naked eye as a faint patch about as large as the Moon. It shows up well in binoculars and should have a generally triangular shape. **M48** is 2,000 light years away.

Comets

One of my favourite quotations is by the comet hunter David H Levy: "Comets are like cats: they have tails and they do precisely what they want." The history of comets is fascinating and worth getting books from libraries to read up on. From harbingers of doom to kings and rulers, they are now harbingers of possible doom to Earth. In the meantime, enjoy the sight of these dirty snowballs and their capricious tails.

Leo – the Lion (March to May)
(Northerly aspect)

Strangely enough, this constellation actually looks like a crouching lion – if you are standing on your head. To us, its maned head looks more like a large sickle with its handle upwards. It is very easy to spot.

Though Leo has a nice collection of spiral galaxies at magnitudes 9 and 10, they are tantalisingly beyond the reach of all but the largest binoculars. To compensate, it has some attractive multiple stars.

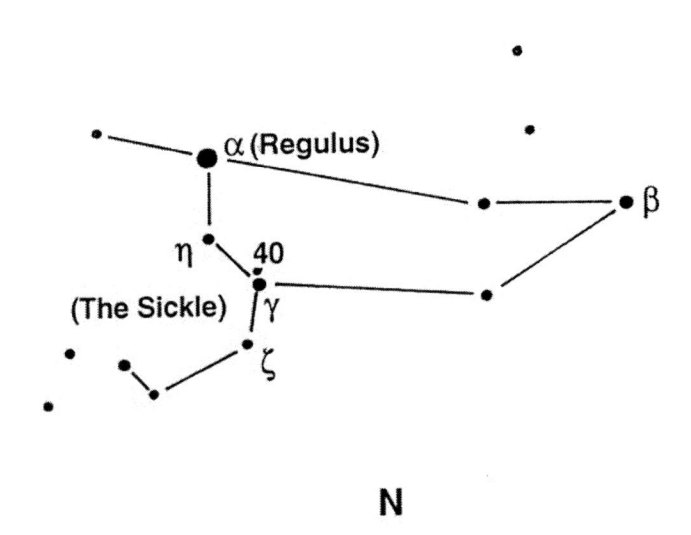

N

α **Leonis (Regulus)**. The brightest star in Leo at the top of the sickle's handle. Also dubbed 'The Lion's Heart' it is a blue-white star only 77 light years away and a bright mag. 1.4. It is a binary, with an orange-red mag. 7.7 companion 3' away, so it is easily split in binoculars, though faint. Regulus is of extra interest as it lies close to the Ecliptic, so it is often in conjunction with the outer planets and will occasionally be occulted by the Moon. It is worth keeping an eye on.

Move down (north) from Regulus, past the first star (η) and east to find γ **(gamma) Leonis**, a 3rd mag. yellow star. It is actually a beautiful binary of a gold and a yellow star but binoculars cannot split these. What you will see is their wide optical companion, a mag. 4.8 star called **40 Leonis**.

Now move 3° directly below γ down the sickle to its bottom eastern star. That's ζ **(Zeta) Leonis**, an optical double, comprising a giant white star (mag. 3.4) and an orange mag. 5.9 star, separated by about 5'. Slightly further away is another mag. 5.8 star. An optical triplet!

Lepus – the Hare (November to April)
(Overhead aspect)

Lepus's stars are not very bright but still easily seen. They seem to form a distinct pattern. The Greeks saw it as a hare, or rabbit always under Orion's feet and chased by his dogs, Canis Major and Canis Minor. They were looking at it upside-down from the way we see Lepus. When facing north and joining the dots, I see a distinct shape of a man running from right to left, his arms flung out, his front knee and leg raised and his other leg pushing hard from behind. But then, I suppose it's all in how you join the dots.

Though it has many objects of interest to telescopes, there is only one binocular object.

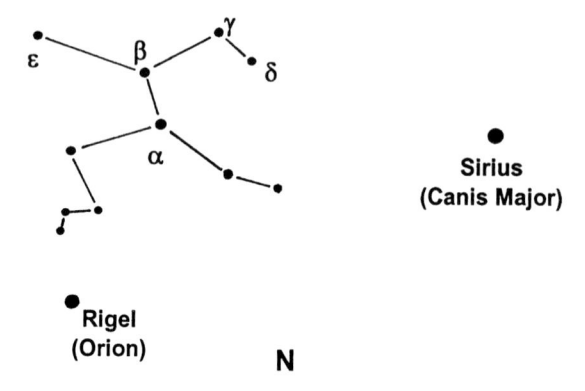

Lepus is directly 'beneath' Orion's feet, so we see it 'above' or south of the handle of the Saucepan and Rigel. When you find the line between its two brightest stars, α and β **Leporis**, γ **Leporis** is just 4° to the right of β.

γ is a double star (separation 1.5') consisting of a mag. 3.6 yellow and mag. 6.2 orange, about 29 light years away. The second star is visible but close and fainter so hold your binoculars steady. Though not a true binary (they don't orbit each other), they are close and travelling through space together (i.e. not just an optical double).

How many stars are there in the Universe?
How long is a piece of string? The approximate number of stars can be calculated as follows. An 'average' galaxy has about 300 billion stars. There are an estimated 100 billion galaxies in the Universe (possibly even more). This gives a figure of 30,000 billion billion stars, or 30 with 21 zeroes after it. **Wow!**

Libra – The Scales (April to August)
(Overhead, slightly northerly)

One of the Zodiac constellations, this can be found hovering over the head of Scorpius. In fact, in earlier times, these faint stars formed the scorpion's claws. Now they represent the scales of justice, though the blindfolded lady holding them is in the adjoining constellation Virgo.

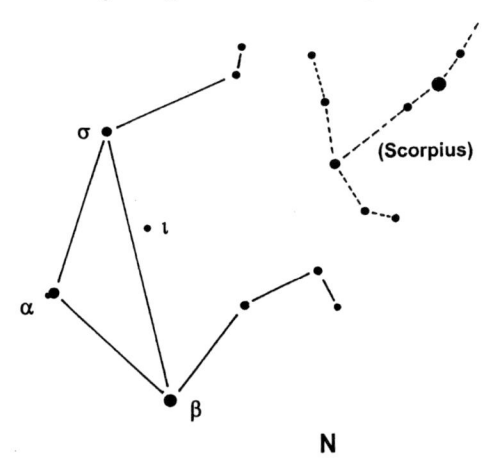

If you draw a line from Antares in nearby Scorpius through the centre of the scorpion's head, this will point you towards α **Librae**, about 16° from the scorpion's head. α Librae (its name mean 'Southern Claw') is a binary star very suitable for binoculars. Its primary star is a mag. 2.8 blue-white, while its mag. 5.2 white companion is a wide 3.9' apart. They are 77 light years away from us. This means they are about 1/10th of a light year apart. That's very close for stars. But to complicate matters, the primary is itself a very close double star, unable to be split by the largest telescopes.

Move below and right of α by 9° and you will find another bright star, β **Librae**.

β **Librae** (its name means 'Northern Claw') is celebrated for a most unusual reason – its colour. While a mag. 2.6, some claim that it has a 'greenish tinge', like a pale emerald. Many dispute this as it is a single star and these are not known to be green. But as many claim it is green, it has the disputed distinction of being the only known green single star. You look and decide…is it greenish?

Now move upwards (south) from β by 11° and left (west) by about 1°. You should see a faint 4.5 mag. white star called ι **(iota) Librae**, about 370 l.y. away. Close to it, half a Moon diameter away, will be a 6th magnitude white companion called **25 Librae**. This is much closer at 220 l.y. This is all binoculars will reveal, while in fact ι **Librae** is a double-double star, making it a very complex system.

Lupus – the Wolf (April to September)
(Southerly aspect)

Lupus is a large constellation made up of 2nd and 3rd magnitude stars, many of which are telescope doubles. There are also plenty of optical binocular doubles. It can be found in-between Scorpius and the Two Pointers of Centaurus. It contains an open cluster and a globular cluster, suitable for binoculars, though the globular will be a challenge.

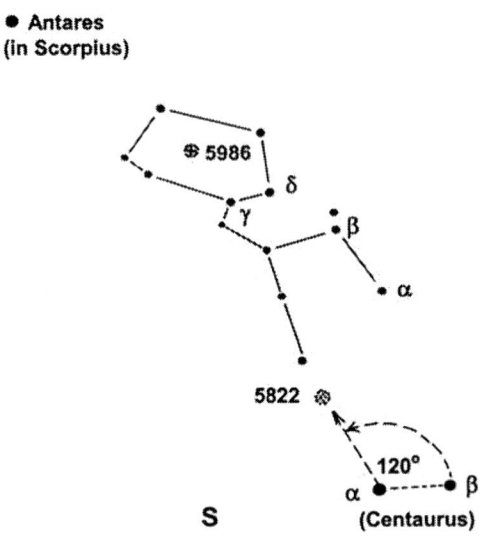

If you rotate the line from α to β Centauri anticlockwise about α by 120°, and extend it by half, you should rest on an open cluster **NGC5822.** This mag. 6.5 cluster is very large (almost 1° across), and has about 150 loosely arranged stars from mag. 9 and fainter. It looks like a sprinkling of sand on black paper.

NGC5986 is a globular cluster found about 14° from Antares in the centre of the wolf's head. (See map for exact location.) This 8th mag. globular, visible only to larger binoculars, is 33,000 light years away and appears as a faint round patch. Try this from a dark clear site as it is a difficult binocular object.

Lupus is also lovely to explore for its many optical binocular doubles. Take the time, with your binoculars steadied, to tour the various stars in the head, neck, shoulder and legs of the wolf. Many of these stars form lovely pairings or triplings of faint and not so faint stars, some wide, others close, just needing binoculars to bring them out. There are too many to list here – just scan the area and enjoy them. See how many groupings you can count.

Lyra – The Lyre   (July to September)
(Low northerly aspect)

Representing a mythical lyre, this constellation is easily found after you locate its main star Vega, the 5th brightest star in the sky. Above and to the right of Vega you will see a very distinctive and regularly shaped 4° long parallelogram. Though you cannot see it in binoculars, between the two top stars lies a famous planetary nebula, M57 the Ring Nebula, which can be spotted as a small 'smoke ring' in even a modest telescope.

Find Vega, then 1.5° (3 moon diameters) north-east you will find ε **Lyrae**, a famous double-double star. Your binoculars will easily split it into a pair of 4.6 mag. white stars but a modest sized telescope will reveal that each of these is a binary star. Four for the price of one.

Now move to the bottom left star of the parallelogram, ζ **Lyrae**. This is a close optical double of mags. 4.4 and 5.7 that binoculars can just split. My 12x50s show a pretty but close pair. They may be too close for 7x binoculars. Steady your binoculars and see if you can split it.

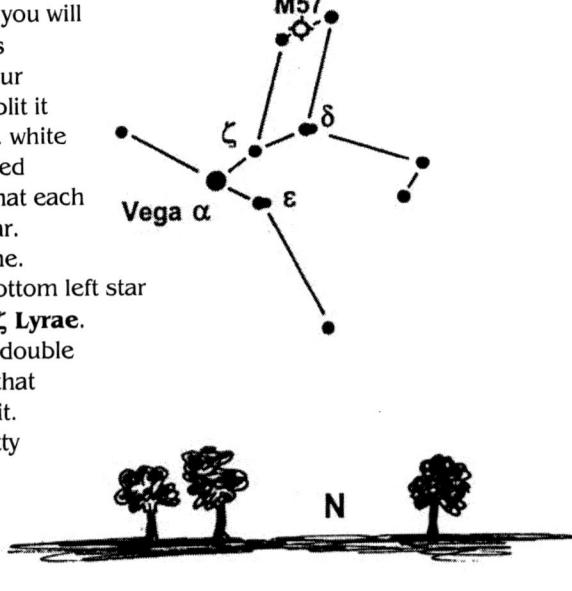

Now move right about 2° to the parallelogram's bottom right star. δ **Lyrae** consists of a red giant (mag. 4.2) and a blue-white star (mag. 5.6) making an attractive optical double. It should also be visible as a double to your naked eye.

There is a very famous object between the two top stars of the parallelogram. This is **M57**, the **Ring Nebula**. This is the quintessential planetary nebula, only 2,000 light years away. In a small telescope, it looks like a smoke ring or a doughnut, with the hole in the middle plainly visible. However, while some claim to be able to see it in binoculars, others cannot. If your binoculars are large enough and the sky is good enough for you to see it, the most you will see would be an elliptical misty patch. Consider that a great achievement, then hunt down a friend with a telescope to see it even better.

Monoceros – The Unicorn (January to April)
(High northerly aspect)

It may have only one horn, but this faint (but large) constellation found between Sirius, Orion and Procyon (of Canis Minor) has a multitude of binocular objects. No stars this time, all NGCs. But first you'll need a reasonably dark sky to find the constellation.

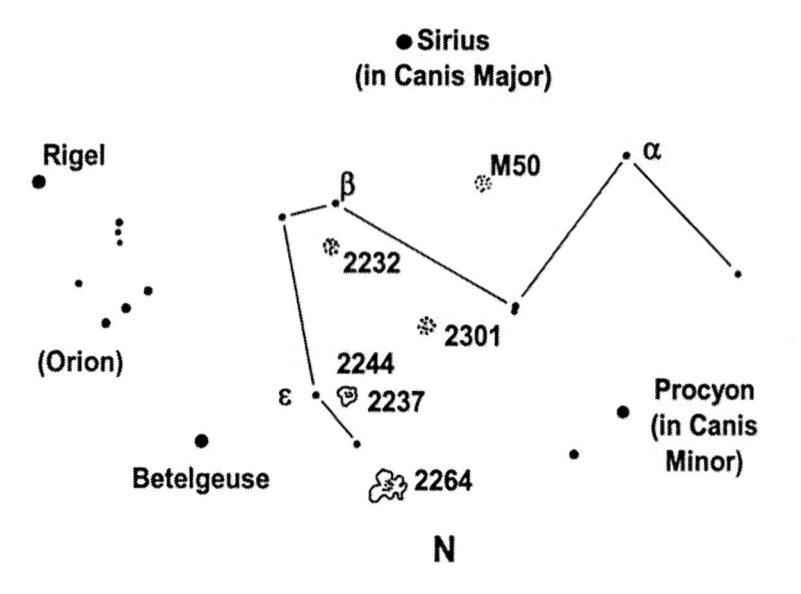

Find Sirius and draw a line to Procyon. The open cluster **M50** can be found approximately 1/3 of the way from Sirius and is easily observed in binoculars. 3,300 light years away, this cluster has at least 80 stars in it, covering an area about 1/2 that of the Moon. Some say they can see a heart shape in it.

Now find Rigel (white) and Betelgeuse (red) in Orion. Using them as one side of an equilateral triangle, imagine the other triangle point in Monoceros. There you should find a 5th mag. star – that's β Mon. About 2° below (north) of β Mon. you will find **NGC2232**, a very loose Open Cluster of about 20 stars 1,200 light years away, covering an area about that of the Moon.

Now move 10° down (north) from 2232 to a 5th mag. star ε Mon, then 2° right (east). This brings you to **NGC2244,** a star cluster created from and at the centre of the very famous Rosette nebula **NGC2237**. Binoculars can easily spot the cluster with its six main stars forming a rough, bent rectangle about half a Moon diameter long. The Rosette itself is a very complex nebula over 2 Moons in diameter. On a dark night with clear sky, you may just be able to 'see' it as a soft haze surrounding the cluster but it is very illusive. Another way of finding NGC2244 is to follow the line from Betelgeuse to Procyon about 1/3 of the way. Both cluster and nebula are about 5,000 l.y. away.

If you take the same line between Betelgeuse and Procyon, 4° below the mid-point you'll find **NGC2264**. It is another example of being able to see the star cluster but not the nebula that surrounds it, which is a pity as the Cone Nebula is particularly beautiful. The cluster of about 40 stars appears, when seen from our southern sky, to have a shape like a Christmas tree about the size of our Moon, with its brightest star 5th mag. 15 Mon. at its base. It is 2,600 l.y. away.

Now, if you find the centre of the triangle formed by Betelgeuse, Sirius and Procyon, **NGC2301** will be a little off centre towards Procyon. NGC2301 is a nice open cluster of 60 to 80 stars, 2,500 light years away. It has a distinctive shape like a straight chain with straggling offshoots.

Discovery of Uranus

A German expatriate, William Herschel, had moved to England with his sister, Caroline. He made his living as a musician, music teacher and church organist, but his hobby was astronomy. He set out to make his own telescope and in the process made about 200. He also set out to map the entire heavens – no small ambition in his spare time.

One night in 1781, during a routine survey, he came across a strange star. He checked other charts and could not find it listed. He came back to it night after night and decided it had moved fractionally. He announced his find and was credited with discovering the 7th planet, which was named Uranus.

The King of England gave him a handsome pension and Herschel gave up music and became a professional, and very productive, astronomer.

In one of those intriguing "what ifs" of life, Herschel made so many discoveries in astronomy, one wonders how far modern astronomy would be set back today "if" Herschel hadn't happened upon Uranus that night.

Cosmic Coincidence:

William Herschel lived from 15th November 1738 to 25th August 1822, a life span of 83.7 years. The planet he discovered, Uranus, has an orbital period (year) of 84 years. Coincidence, yes, but apt.

Musca – the Fly (February to August)
(Southerly aspect)

To find Musca, simply locate the Southern Cross. Musca hangs directly 'below' it, that is, at the foot of the Cross. It originally began its life as Apis the Bee, but for some reason (possibly because it was being confused with Apus) later became the Fly. (Drat!)

Musca has a distinctive quadrilateral shape, formed by 3rd and 4th magnitude stars α, β, γ and δ Muscae. They are moderately faint but easy to find on a dark night.

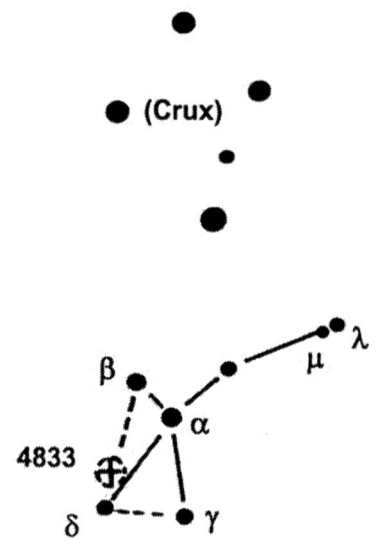

At the bottom left hand (eastern) corner of the quadrilateral, what looks like a single star δ **Muscae** is actually a mag. 3.6 orange giant about 90 l.y. away, with a nice blue-white 6th mag. optical companion (wide 5' separation) over 2,000 l.y. away. That's a big difference.

At the opposite end of the fly, what looks like one star binoculars reveal as a wide optical double (λ and μ) with an attractive 5th magnitude orange and 4th magnitude blue-white contrast.

Less than 1° above the bottom left (east) star δ Muscae is **NGC4833**, a 7th mag. globular cluster visible in binoculars as a faint smudge, despite its distance of 18,000 light years. Add this one to your globular collection.

Norma – the Set Square (May to September)
(Southerly aspect)

This small and totally insignificant constellation of 4th magnitude stars, representing a carpenter's level or a draughtsman's set square is tucked midway between the Pointer stars and Scorpius. It is small, its 'set square' only 4.5° in length. However, one consolation is it lies smack in the rich fields of the Milky Way. Frankly, Norma is tricky to locate in the sky since its stars are so few and faint. Persevere from the map below and you will find it.

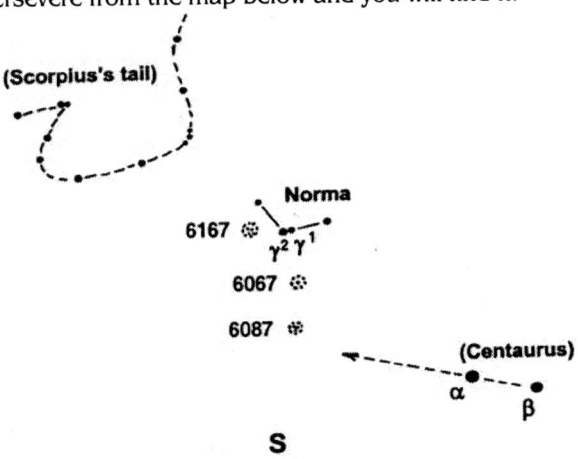

If you extend the line from β **Centauri** through α **Centauri** for 3 lengths, it will end very close to the open cluster **NGC6087**. Or, if you can locate γ **Normae** from the map, drop down about 8° (1.3 Southern Cross lengths) to find it. **NGC6087** is an attractive open cluster of about 50 stars containing radiating lines suggesting spinderly legs. In your binoculars it will appear to cover an area about one half the diameter of the Moon.

Halfway between **NGC6087** and γ **Normae** is another open cluster, confusingly called **NGC6067**. This has around 100 stars, also about half a Moon diameter and is also attractive. Though they have similar NGC numbers, they are each different in shape and character.

Then, about 3° east of γ **Normae**, you will find a faint hazy patch of stars. This is **NGC6167**, 1 magnitude fainter than **6067** and **6087**, but also with about 100 stars, though the stars are not easily resolvable.

γ (gamma) **Normae** itself is a wide optical double (they are about one Moon diameter apart). The western-most component, γ^1, is a mag. 5.0 yellow-white supergiant about 1,500 light years away. By comparison its companion on the left, γ^2, is a mag. 4.0 yellow giant only 130 light years away. When you look at this pair, consider how much bigger and brighter the furthest star, γ^1, must be (at over 10 times the distance) to be almost as bright as the closer γ^2.

Ophiuchus – The Serpent Holder   (June to September)
(Northerly aspect)

This is a very large constellation, with all of its brighter stars around the perimeter. It doesn't stand out as an obvious constellation but it is of immense astronomical interest, even for binoculars.

The easiest way to locate Ophiuchus is to face north with Sagittarius overhead on your right and Scorpius on your left. Ophiuchus will be that area of the sky immediately between and to the north of Sagittarius and Scorpius. It covers an area of about 40° (north-south) and 30° (east-west). Its main stars range in magnitude from 2.1 to 3.8. That's why they don't stand out as much as, say Antares at mag. 1.

The perimeter of the constellation, inside which there are very few naked eye stars, has been dubbed 'The Coffin', not to be confused with Job's Coffin in Delphinus.

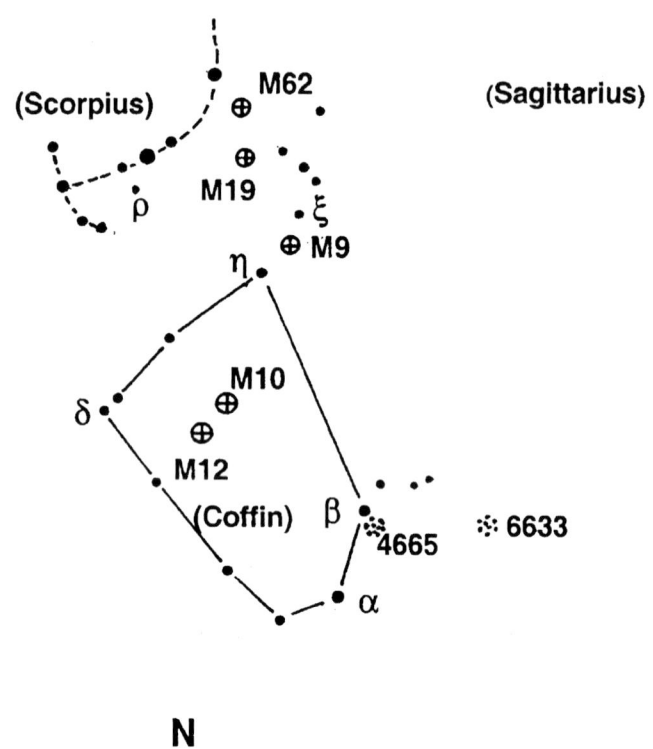

N

First, let's find a multiple star. ρ **(rho) Oph.** is right on the border of Ophiuchus and Scorpius and can be found 3° north of the red giant Antares. The main star of ρ **Oph.** is a 5th mag. binary (just naked eye visible but not splitable by binoculars) but binoculars will reveal companion stars of mags. 6 and 7 flanking it with 2° separations. The three amigos.

Globulars: The main attraction of Ophiuchus is its globulars – it contains 6 Messier numbered globs, at least 3 of which binoculars will reveal. Again, it depends on your binoculars and the darkness of the sky.

M9 is the smallest and faintest at mag. 7.9, but has a very concentrated centre. Of **M10**, **M12** (both mag. 6.6) and **M19** (mag. 7.2), **M19** has the most concentrated centre and is the smallest of the three. **M10** is the most loosely packed and **M12** is of similar size. They are all very far away. **M9** is about 26,000 light years away (which explains its small size), **M10** is 14,000 light years, **M12** is 18,000 light years and **M19** about 20,000 light years.

To locate **M19**, move directly west 7° from Antares (in Scorpius). **M19** is a difficult object for binoculars and a clear dark sky will be needed. It appears as a compact fuzzy patch, like a bright star out of focus. Now move down 11° and eastward and identify η **Oph** (mag. 2.5), then go back up 6° SE to ξ **Oph** (mag. 3.5). Exactly midway between them is **M9**, which is also a difficult object being relatively small. (Don't be surprised if you can't nail **M9** and **M19** – binoculars larger than 7x50 will be required.) A helpful marker for M9 and M19 is the 4° long chain of 3rd magnitude stars that span a rough arc between them. The ends of this 'arc' give the approximate locations of M9 and M19.

More easily seen are **M10** and **M12**. Locate the mag. 2.7 star δ Oph on the far left. Move 10° due east to find **M10**. Then move down and west for 3.5° to find **M12**. This means you should see them both in the same binocular field of view. This allows you to compare their differences of structure.

Another globular cluster in Ophiuchus is **M62**. However, some references would have it in Scorpius as it is right on their borders. At mag. 6.6, it should be visible in binoculars in a clear dark sky as a very faint fuzzy star. To find it, imagine the line between Antares (the red giant in Scorpius) and **M7** (the open cluster near Scorpius's tail). **M62** lies 8° or about 1/3 the distance from Antares along that line.

But there's more…

There are two open clusters that show up well in binoculars. **NGC6633** contains over 40 stars, covering an area just less than the Moon. To the naked eye it appears as a faint 5th mag. patch of light on the eastern outskirts of Ophiuchus. If you can identify α and β **Oph**. from the map (at the bottom right), swing the line from α to β anticlockwise around β by 90° , extend it by 3° and it should land on **NGC6633**.

IC4665 is found only 1° north of (down from) β **Oph**., just east off the line from β to α. This open cluster covers a large area, almost twice that of the Moon, but the stars are very scattered, there being only about 25 of them. This cluster is on the limit of naked eye visibility but shows up well in binoculars.

Orion – The Hunter (December to March)
(High Northerly aspect)

Appearing in every ancient culture's folk lore as either a great hunter or warrior, Orion is very easy to spot. You look up and there it is, the three bright stars of his belt, the sword (or dagger), and the bright white Rigel above with blood red Betelgeuse below. This is major *'Wow!'* territory. Of course, to us Down Under, we see it up-side-down and it appears as a large saucepan, not a club wielding hunter and is probably easier to spot as such.

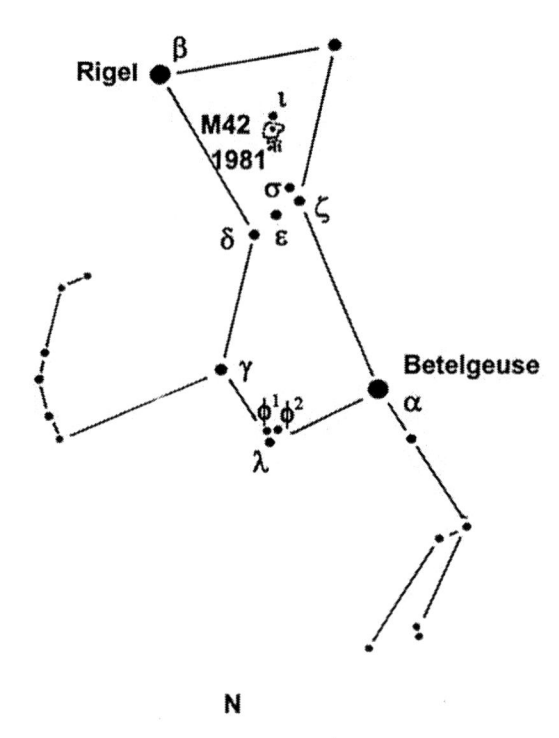

N

α **Orionis** (below the belt) is the famed Betelgeuse (pronounced bet-el-gerse). It is a fantastic star of unimaginable proportions. A red supergiant 427 light years away, its diameter is estimated at 400 to 500 times our Sun's. That means if put at the centre of our solar system, its outer surface would extend beyond the orbit of Mars. **Wow!** But imagine this. Its surface temperature is about 3,000°C, but the average density of Betelgeuse is about 1/10,000th of the air you are breathing. No wonder they sometimes call it a 'red hot vacuum'. It just looks like a hot red star in binoculars but think on those facts as you gaze upon it.

β **Orionis**, above and to left of the belt, is Rigel, 770 light years away, the 7th brightest star in the sky (mag. 0.1) and as a white hot supergiant giving out more than 57,000 times the light as our Sun, it is one of the brightest objects known in our galaxy. Rigel and Betelgeuse couldn't be more opposite in colour.

Orion's Belt (the base of the Saucepan) comprises δ, ε and ζ Orionis. Both δ and ζ are blue-whites and multiple stars, while ε in the centre is a single blue star. It is just above ζ that one could see the famous Horse Head Nebula if you had a big enough telescope. It is way out of a binoculars' league.

δ Orionis is the left hand star in the belt. A complex multiple star at mag. 2.2, your binoculars may reveal an optical double, the unrelated star being mag. 6.9. As their separation is only 52" and there is a big difference in magnitudes, larger magnifications (say 10x or 12x) and steady binoculars are best.

Sitting above ζ Orionis, the right hand star in the belt, is **σ Orionis**, a tantalising mag. 4 multiple star. While its full appearance as a mini-solar system of seven stars needs a moderately powerful telescope, one of its mag. 6 components, at 3' separation, can be split with binoculars but it is very faint.

A special treat in Orion is **'the Sword of Orion'**. You can't miss it, it's the bit that resembles the handle of a saucepan. To the naked eye it looks like a ragged string of three or four fuzzy stars, poised above the three bright stars forming the 'belt'. But with binoculars, a whole new world opens up. Contained within the same field of view, you will see a chain of wonders.

The show stopper in the middle is the **Great Nebula of Orion, M42**, a huge glowing nebula 1,500 light years away surrounding the 'star' Theta Orionis, which is really a multiple star system. This nebula is 15 light years in diameter and is giving birth to myriads of stars even as you watch it. This is one of the most beautiful and studied objects in the heavens and you will come back to this time and again. Try and find a friend with a telescope to study it closer.

But that's not all. Below **M42** another glowing nebula surrounds a tight knot of stars, and below that again is a sprinkling of about twenty stars in open cluster formation **(NGC1981)**. And above **M42**, completing the chain, is the bright star Iota Orionis, with some friends. Each of these is beautiful on its own, but to see them together as in a bank queue, in one glimpse, is marvellous. So do whatever you can to keep your binoculars steady and you will gaze at this chain of stars and gases for ages.

Finally, after the sublime, here's a pretty little sight.

Find λ (lambda) Orionis – it's effectively the head of Orion – and you will see a knot of three stars with your naked eye. This little group is called **Collinder 169 (or Col169)**. With your binoculars, you can bring them closer to show a distinct triangle comprising λ, ϕ^1 and ϕ^2. And between λ and ϕ^1 you will see a pretty little string of three faint stars, like a chain of stepping stones. No astronomical marvel, but pretty all the same.

Pavo – the Peacock (June to November)
(High southerly aspect – but circumpolar)

This constellation was introduced in the 16th century by Dutch navigators, perhaps to add to the growing menagerie in the sky. The peacock, of course, is that beautiful bird with a tail of 100 eyes. It was a sacred bird to the Greek gods and played a part in one of their many convoluted tales. The main stars of this constellation are moderately faint at mag. 3, with the exception of α **Pavonis** at mag. 1.9. Pavo can be found about halfway between Corona Australis and the South Celestial Pole.

The only binocular object of interest is a magnificent one, the Globular Cluster **NGC6752**. This can be found about 10° west of and 3° below α Pavonis. If this had been visible from France, it certainly would have a Messier number. Arguably the third largest (in apparent size, about ½ a Moon diameter) of the globulars, and at 5th magnitude, the 7th brightest in the sky. Only 15,000 light years away, it is a treat for binoculars. It has a fairly concentrated center with an even distribution of outer stars, with a 'bright' foreground 7th mag. star on its fringe.

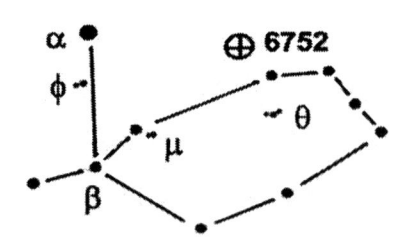

It is also worth scanning the area of this constellation for more of those pretty patterns and groupings of fainter stars. Look in the areas near φ (phi), μ (mu) and θ (theta).

θ is in a group of 5th and 6th magnitude stars about 1° long, with a mixture of colours from white to red.

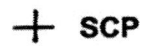

φ is a wide 5th mag. yellow double, forming a triangle 1° long with ρ, another 5th mag. yellow star.

μ contains a closer double of 5th and 6th mag. red stars, only 8' wide.
And there are plenty more.

Pegasus – the Winged Horse (September to November)
(Low northerly aspect)

In Greek mythology, Perseus slew the Gorgon Medusa, and when blood from her neck struck the ocean waters, Poseidon caused it to give birth to a son, Pegasus. The most identifiable feature of this famed flying horse is 'the Great Square of Pegasus' which represents the horse's torso. The sky in that area is strangely lacking in other faint stars, which may be why Pegasus' faint 2nd and 3rd magnitude stars are able to be identified. It can be found about 30° to the east of the 'kite' shaped Delphinus.

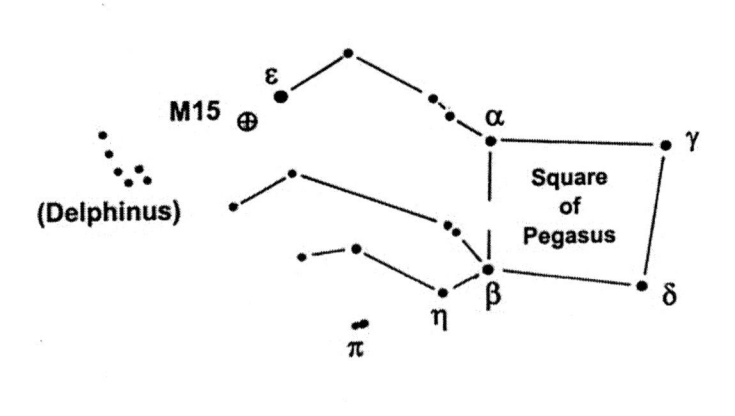

The bottom right hand star of the Square, δ **Pegasi**, actually leads a double life. It is also called α Andromedae and represents the head of the maiden Andomeda.

Find β **Peg** at the bottom left of the Square. Move down and left about 4° to find η **Peg** (mag. 3). Continue another 9° and you will find π **Pegasus**, a wide (9' separation) optical double comprising faint white and yellow giants. Imagine 250 and 280 light years.

M15 is the main binocular feature of Pegasus. Try and locate the yellow 2nd mag. star ε **Pegasus** from the map, about 16° east and above the kite in Delphinus. Move back 4° NW and you will easily locate this rich and compact **globular cluster**. It can be recognised as a brighter fuzzy 'star' inside a triangle of stars. At mag. 6.4, it's just beyond naked eye visibility – it is 33,000 light years away, after all. Not as spectacular as **NGC6752** in **Pavo**, but still impressive.

Perseus
(December to January)

(Low northerly aspect)

The constellation Perseus is mythologically entwined with Andromeda, Cetus, Pegasus, Cepheus and Cassiopeia (the latter two are not visible to most southern observers). Perseus was a young hero who fought and slew the Gorgon, Medusa, rode on Pegasus' back and rescued the fair Andromeda from the monster Cetus. He wedded Andromeda, then killed her treacherous parents. Eventually, he reputedly fathered the nation of Persia. A busy young man.

It has a number of attractive binocular objects but is very close to the northern horizon so viewability is a question of latitude and a clear horizon. Its magnitude 2 to 4 stars can be found midway between those of Andromeda and Auriga and directly beneath the Pleiades cluster in Taurus.

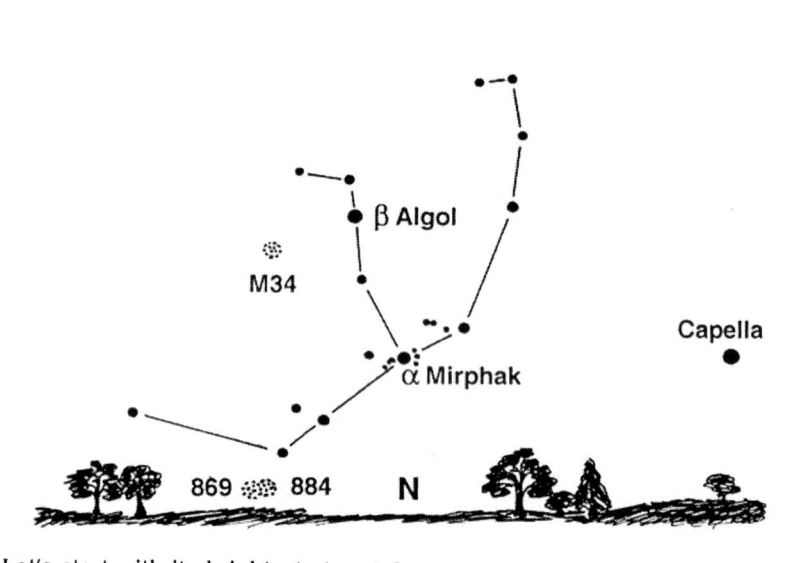

Let's start with its brightest star, 1.8 mag. α **Persei**. α, or Mirphak, is a yellow-white supergiant 590 light years away. What is of real interest for binoculars is the delightful cluster of stars in its vicinity, named Melotte 20. It covers a largish area of sky, over 5 moon diameters and contains some distinctive chains of stars, most of them of 5th and 6th magnitude. It appears to me as an elongated shape, like a cone or triangle, with Mirphak at its apex.

About 10° above (south) of α is β **Persei (Algol)**, a very famous star which was the first discovery of a special class of star which now bears Algol's name. Algol means 'the demon' and represents the severed head of Medusa, the Gorgon. (It's safe now to stare at this star – you won't be turned to stone.) What is special about it is that it belongs to a class of variable star called an 'eclipsing binary' where the star's variable brightness is due to an invisible companion star that partially eclipses our view of it, not due to an inherent variability of the star itself. You can observe this phenomenon yourself as its period of variability is very short. It happens every 2 days 20 hours 53 minutes. Normally mag. 2.1, when the eclipse occurs it drops over 1 magnitude to 3.4 before returning to 2.1 about 10 hours later. If you have some patience over a period of 3 or 4 days, you should be able to observe the moment when Algol is either eclipsed or moves out from eclipse – quite a dramatic experience.

Move from β **Persei** about 4° west and 1° north (down) and you will find **M34**, a fairly loose open cluster with about 60 stars occupying an area similar to a full moon. Though it is 1,500 light years away, you should be able to resolve its stars in your binoculars.

The following pair of objects, visible to the naked eye and absolutely glorious in binoculars, is unfortunately barely over the horizon from New South Wales latitudes, having a declination of +57°. However, from latitudes in the northern states, it should be viewable, though still close to the horizon. I'm referring to the **Double Cluster in Perseus**, **NGC 869** and **884**. While each cluster is about the same area of the moon, the easternmost **884** is more loosely scattered than the richer and more compact **869**.

Neptune found

Based on calculations using Newton's laws of gravitation, something was not right with the observed orbit of Uranus. Uranus was not being found where calculations said it should be. People were beginning to suspect flaws in Newton's theory, but two astronomers continued to trust the theory and so proposed that there was a planet further out than Uranus that was perturbing its orbit. An undergraduate in Cambridge, John Adams and a professional French astronomer Urbain Leverrier, working independently, used some very heavy number crunching to calculate possible locations of such an 8th planet.

Eventually such a planet, Neptune, was found where predicted by both astronomers in 1846. Both were credited with the find – this led to much Anglo-Franco angst.

Pisces – The Fishes (September to December)
(Low northerly aspect)

This old zodiac constellation, containing two fish tied at the tail (it supposedly represents Venus and her son Cupid who changed into fish when diving into the Nile to escape the monster Typhon, then tied their tails together to avoid being separated), is large and faint, its brightest star (η, not α) being only mag. 3.6. It will take some patience to trace it out in the sky from the map. However, it contains some double stars of interest. A guide to finding Pisces is that it lies above (south) and along the right (east) side of the Square of Pegasus. Also, you should see the circular feature (called, strangely, the Circlet) of seven stars that represents the head of one of the two fish.

Find the top (southern) side of the Square of Pegasus. Move about 1/3 the way in from the western star, then move up (south) about 13° to the top of the Circlet. You will find a white mag. 4.9 star κ **(Kappa) Piscium** which is 160 light years away. It has a yellow optical companion at mag. 6.3, invisible to the naked eye but clear in binoculars 9' from κ. 5° east and 2° north of κ is a faint 5th mag. deep red star called **19 Pisc**. This varies its brightness between mag. 4.8 and 5.2 at irregular intervals. This half magnitude difference can be detected, particularly in binoculars, if you watch it over a long period of nights.

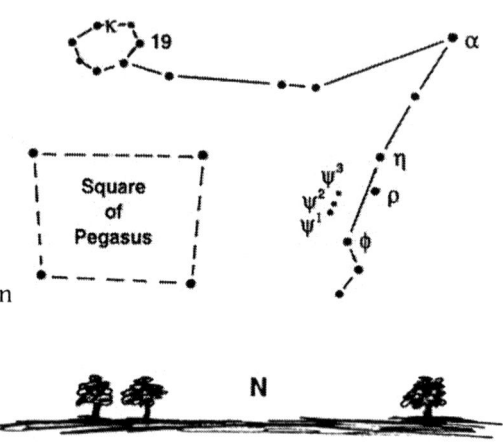

Now find the right (eastern) side of the Square and come down (north) 1/3 of the way from the top star. Then move 17° right (east) to find a 5th mag. white star ρ **(Rho) Piscium**. This is a fine optical double (7' apart) comprising a mag. 5.4 white star (85 light years away) and a mag. 5.5 orange giant (310 light years away).

Forming a straight line parallel with the line between ϕ and ρ **Piscium** is a row of three stars called $\psi^{1,2,3}$ **(psi) Piscium**. It's strange they have the same letter identification as they are so far apart (1° each), but they form a pretty asterism of blue, white and yellow stars of equal magnitudes and distances of 230, 114 and 98 light years respectively.

Puppis – the Stern (January to May)
(High southerly aspect)

Puppis is part of the trilogy of constellations making up the ship Argo. (The other two parts are Carina and Vela.) It can be found mid-way between Sirius and the False Cross, butting up against Carina and Vela. Being in the rich star fields of the Milky Way, it offers a multitude of stars and star clusters. Use the map and angles between key stars to locate the objects.

First a double star, ξ **(xi) Puppis**. Find the base of the dog's tail in Canis Major and move 10° east to find it. This mag. 3.4 yellow supergiant forms an optical double (4' separation) with a mag. 5.3 orange giant. ξ is a good marker for **M93**, which lies just 1.5° away in the direction of Sirius. It's a border-line naked eye cluster, 3,600 light years away, of about 80 stars with a suggestion of an arrow head shape.

10° North of M93 and 12° due east of Sirius are two clusters about 1.5° apart, each of them about the area of the Moon.

M46 (on the east side) is a large sprinkling of over 100 very faint stars, 5,200 light years away. Binoculars will not resolve these stars, but it will look like a very faint filmy patch. (Obviously much better on a dark night.)

The nearby **M47** has about 30 stars, but these are brighter and more distinct.

Then, moving down towards the False Cross, about half way you will find the star ζ **(zeta) Puppis** (mag. 2.2 and one of the hottest stars known in our galaxy). From ζ move about 3° back towards Sirius to find the star **b Puppis** and **NGC 2477** nestled against it. In fact ζ, **a** and **b Puppis** form a nice triangle and help to find **2477**. This open cluster with over 160 stars and border-line naked eye visibility 4,000 light years away has an uncanny resemblance to a less compact globular cluster. It is very striking in appearance and it has been suggested that if it had been further north, it would have a Messier number.

From **2477**, move 1.5° NW and you will find another open cluster, **NGC2451**, which is scattered around the 4th mag. orange star **c Puppis**. You should be able to pick out over a dozen distinct stars in this group.

Now locating the ζ, **a** and **b** triangle again, and moving 3° east from ζ in the direction shown on the map, you will find **NGC2546**. Though identified as a separate cluster, it looks like a part of a rich star field. However, within this star field, it seems to form a distinct group of brighter stars. Very pretty.

Pluto – Planet or KBO?

Pluto is an enigma, the whipping boy of planetary astronomers. Using the same 'perturbation' process as for Neptune's discovery, Clyde Tombaugh, working from Lowell Observatory, predicted a position in the sky and, eventually, discovered Pluto in 1930. The 9th planet, as proclaimed by all astronomy and high school texts.

However, there is reason to believe that if Pluto hadn't been discovered in 1930 but today, it would not be classed as a planet but a Kuiper Belt Object, one of thousands believed to inhabit space from Neptune to beyond Pluto. Many large KBOs have been discovered, with one found in 2005 believed to be 1.5 times larger than Pluto and others only slightly smaller than Pluto.

To add insult to sacrilege, recent computer studies show that Pluto could not possibly have caused the so-called perturbations of Neptune. It was pure chance that Pluto was passing through the area of sky that Tombaugh had calculated the 9th planet to be in. Ouch!

Sagitta – the Arrow (July to October)
(Medium high northerly aspect)

A very small constellation with faint 3rd and 4th magnitude stars but a delight to observe in binoculars as it is only 5° long and should fit snugly within a binocular's field of view. As its name suggests, it has a distinctive shape like an arrow or dart. In Greek mythology, this is the arrow shot by Hercules to kill the eagle Aquila who'd been feasting on Prometheus' liver. You'll find it below (to the north) of Altair in Aquila, and to the west of the 'kite' in Delphinus. While Sagitta itself holds only one binocular object, it is extremely useful as a starting block for a much more interesting neighbour, Vulpecula.

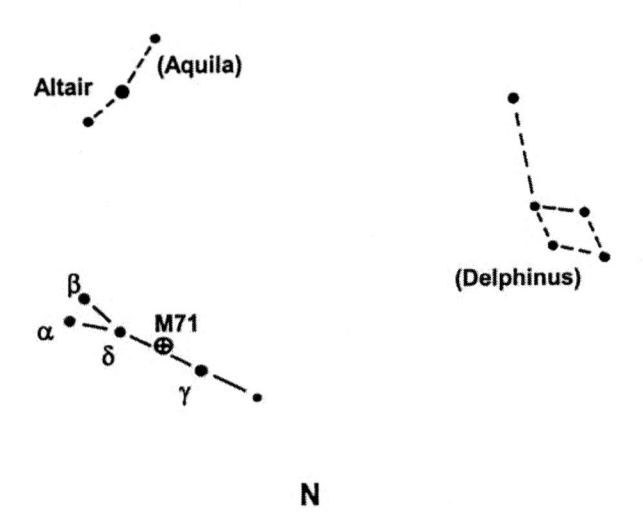

α and β **Sagittae**, while looking like two unrelated stars are actually an optical double of very close stars. They are 473 and 467 light years away, and because of the small angle of separation, a bit more than 6 light years in difference from us. In fact, they are just less than 8 light years from each other. Pretty close as stars go. (Remember, there are only 4 stars closer than that to our Sun.) It always strikes me as odd that many stars that look close together actually are.

Sagitta also offers a challenge, needing a clear dark sky. The challenge is observing **M71**, a faint 8th mag. globular cluster, nearing the limit of binocular visibility. It is located a bit over a 1/3rd of the way between δ and γ Sagittae. As it is 13,000 light years away, it should appear only as a faint, oval shaped misty patch. Averted vision would be useful. Seize the challenge. Obviously, larger binoculars have a better chance.

Sagittarius – the Archer (June to October)
(High overhead)

While the mythological object is a centaur (half man-half beast) holding a bow and arrow, it is more popularly seen as a large Tea Pot with a triangular lid, a triangular spout and a rhomboid handle. It is most easily found due east of Scorpius.

For binoculars, Sagittarius is one of the major **'Wow!'** constellations. Apart from its many objects of interest, the spout of the Tea Pot is the direction towards the very centre of our Galaxy. Focus on that and you are looking towards the densest part of the Milky Way – towards the giant Black Hole that lurks there.

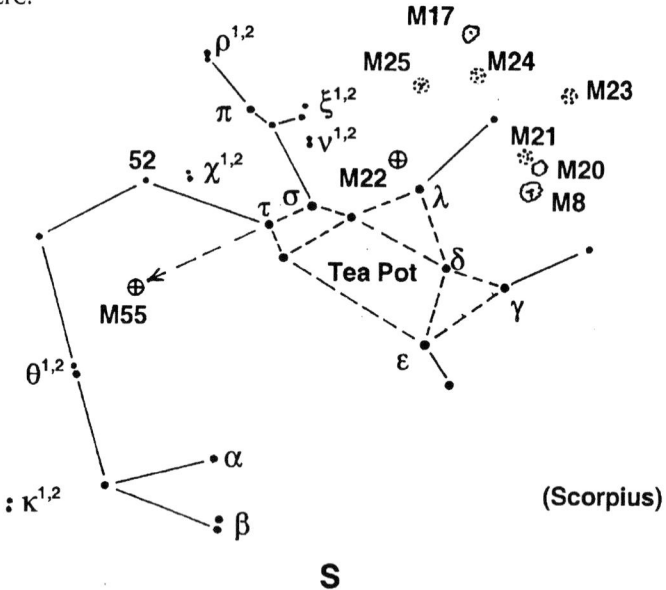

One of the problems with Sagittarius is that there are so many clusters, nebulae and globulars that you don't know where to begin. There are 15 Messier objects alone in Sagittarius, though not all are visible to binoculars.

Let's start with a tour of the **Messiers**.

Start at λ **Sag**. (the tip of the Tea Pot lid). Move 2° to the NE and you will find **M22**, a glorious globular cluster, probably the 3rd brightest in the sky. It is just visible to the naked eye as a fuzzy 5th mag. star. When you look at **M22**, remember how tightly packed the stars are in a globular. This is worth a **Wow!**

Then move 5° due north of **M22** to find **M25**, an open cluster of about 50 stars, slightly larger than the Moon in size. Can you spot the bright 6th mag. yellow supergiant in its midst? Then move about 3° to the west of **M25** to find **M24**, an expansive star field that has a speckled appearance in binoculars. There is some doubt amongst astronomers exactly where **M24** begins and ends. It blends in with some of the other Messier objects. Now move north 2° and find **M17**, the so-called 'Horse shoe', 'Omega' or 'Swan' Nebula. It is a misty oval-shape 1.5 moons x 1 moon in size. Next, **M23** is located 5° further west of **M24**. M23 is an open cluster of about 150 very faint stars, but you will need a good dark sky to see it, and sharp eye sight. If you see it, you'll be delighted with the patterns of chains and arcs of stars it contains.

Moving down towards the Tea Pot's spout, there is a Messier object 5° immediately west of the tip of the lid. **M8** is the **Lagoon Nebula**, visible to the naked eye as a glowing patch of light. It is a delightful nebula with glowing gas and dark lanes and, like the Orion Nebula, one you'll enjoy returning to time and again. One dark patch in its centre gives it the name 'Lagoon'. It is just over 5,000 light years away. To the north of **M8** lie two other Messier objects – **M21** (an open cluster of about 50 stars) and **M20**, the famous **Triffid Nebula**. But these two are unfortunately just beyond the reach of binoculars.

All by itself is **M55**, tucked 7° to the east of the Tea Pot's handle. You can find it by following the line of the handle (σ to τ Sag.) by 3 times its length. M55 is a 7th magnitude globular cluster, 19,000 light years away. It doesn't show a strong central core, so appears a bit filmy in binoculars. Don't forget to use averted vision.

There are a swag of optical doubles in Sagittarius and even though naked eye visible and wide in binoculars, are worth viewing to appreciate their colour or magnitude differences. They are all marked on the map and should be easy to find.

Let's start with β **Sagittarii**. β Sag. can be found directly beneath the Tea Pot's base, twice the pot height distance from the base. Even to the naked eye it appears as two 4th magnitude white stars (20' apart) and are wide apart in binoculars. It is only an optical double, with one star 140 light years away and the other a further 380 light years.

Find π **Sag**. and move 2.5° west. You'll find $ξ^1$ and $ξ^2$ **Sag**., a wide (0.5°) double with mag. 3 orange and a mag. 5 white star (248 and 144 light years away). Then 1.5° south of ξ is $ν^1$ and $ν^2$ **Sag.**, a not so wide pair (13' apart) almost the exact same orange colour and 5th magnitude yet a colossal difference in distance (4500 and 42 light years). $ν^1$ must be reaching old age and bloated to be that bright so far away.

Moving SE from π towards 52 Sag. you'll find the wide mag. 5 double (0.5° apart) $χ^1$ and $χ^2$ **Sag**. You should notice the difference in colour between $χ^1$'s white and $χ^2$'s orange. If you then move 6.5° directly north, you'll find another wide (0.5°) double $ρ^1$ and $ρ^2$ **Sag**. They are a mag. 4 yellow-white (91 light years) and a mag. 6 orange (376 light years).

Moving way south, find $θ^1$ and $θ^2$ **Sag**. as shown on the map. Also 0.5° apart, they comprise mags. 4 and 5 blue-white stars 680 and 170 light years away respectively.

Finally, locate $κ^1$ and $κ^2$ **Sag**. to the east of α and β **Sag**. They are also a wide 0.5° apart and comprise near identical magnitude white stars 150 and 390 light years away.

As you can see, Sagittarius is a very busy constellation and should keep you and your binoculars occupied for quite a while.

Scorpius – the Scorpion (April to October)
(High overhead)

One of the most recognisable constellations – it actually looks like a scorpion, or a giant wool bale hook. Scorpius plays a major role in Greek mythology, being the lowly creature sent by Gaia (Goddess of Earth) to sting and kill Orion for his arrogance in boasting he would kill all Earth's creatures (he was a bit drunk at the time). It is loaded with objects suitable for binoculars, a veritable feast. Another major **'Wow!'** constellation, it is worth revisiting time after time.

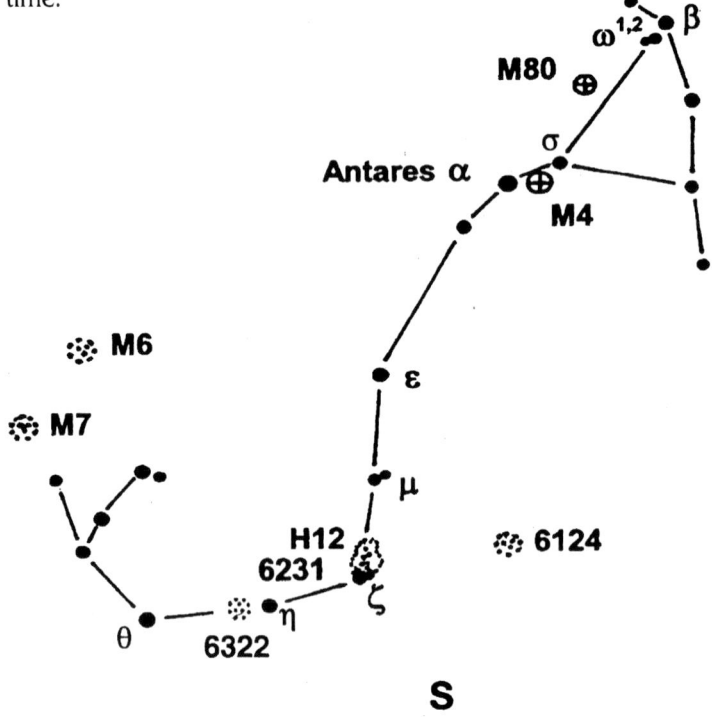

To begin, the bright red star at the centre of the scorpion's spine is α **Scorpii**, or **Antares** (Rival of Mars). That's because when Mars passes close by, it is hard to tell one from the other. Though binoculars only enhance its colour and brightness, it is worth gazing at, knowing that it is a red supergiant star 400 times the diameter of our Sun over 600 light years away. **Wow!** Antares is actually a binary star. Invisible to our binoculars is a fainter blue companion which takes 900 years to orbit Antares.

Down on the base of the tail's curve is ζ **(zeta) Scorpii**, a wide naked eye optical double star. These provide an interesting contrast in colours with the orange and blue-white components, particularly for binoculars. Then, half-way between ζ and ε (back towards Antares) is μ **(mu) Scorpii**, another naked eye double and wide binocular pair of white stars.

Tucked mid-way between Antares and σ Scorpii, and a touch NW is **M4**, a 6th magnitude globular cluster, one of the closest to us at 7,000 light years. It doesn't have a strong concentration, being a loose cluster, and thus is not as easily spotted in brighter skies. In dark skies, however, it is fairly obvious.

At the opposite extreme, at 27,000 light years away and at 7th magnitude, is **M80**, found inline with and exactly midway between Antares and β Scorpii. It is a smaller but much denser globular cluster than **M4** and in a dark sky appears as a small fuzzy spot. This is not an easy one to see. Consider it a challenge.

Easier to see, just 1° below β **Sco.** is the wide optical double ω^1 and ω^2 **Sco.** 14′ apart, its stars, both mag. 4, are a nicely contrasting blue-white giant (424 light years) and a Sun-like yellow star (265 light years).

Two major binocular delights are **M6** and **M7**, both naked eye open clusters located at the end of Scorpius' barbed tail. To the naked eye, they appear as cloudy patches in the sky near the tail. These are a real treat in binoculars and you can make up your own mind as to what shapes or patterns you see in them.

M6 is the one furthest from the tail's barb while **M7** abuts the tail. Some people say they can see a butterfly in M6 which contains about 80 stars, including a 7th magnitude orange giant star. Only 3.5° away towards the barb is **M7**, a larger 3rd magnitude cluster, also with about 80 stars but more spread. Use your imagination. Some say they can see a Christmas tree shape. For most binoculars, **M6** and **M7** should be visible in the same field of view, so admire them side by side.

Right at the bend of Scorpius where the spine becomes the tail, just 0.5° north of ζ, is a cluttered area of star clusters, including **NGC6231**, a cluster of over 100 stars and another large cluster of fainter stars called **H12**. This location in Scorpius is absolutely glorious. In binoculars it gives an impression of two parallel strings of stars. To the side is a clear triangle with two bright blue stars and a bright yellow star. One of the star groupings in this area has been dubbed the 'mini-Pleiades' because of its distinctive shape. If you can steady your binoculars, this area around **NGC6231** and **H12** can be gazed upon for ages.

Then, 6° to the west of **NGC6231** is a very faint misty patch, about ¾ of a Moon's diameter. This is **NGC6124**, a cluster of about 100 mag. 9 and fainter stars. Though its stars are not resolvable in binoculars, it is an enticing object, hanging on the verge of visibility. In a telescope, it is a beautiful collection of sharp stars with blues and reds abounding.

Finally, following the curve of the spine from ζ to η and about 1/3rd the way towards θ, you will find a pretty open cluster called **NGC6322**. It only has around 20 stars but it has an attractive triangular type arrangement reminiscent of the Jewel Box in Crux. It is faint with its brighter stars just resolvable.

Sculptor – the Sculptor (August to January)
(High overhead)

A fairly non-descript and overlooked constellation respresenting, of all things, a sculptor's studio. It is faint (4th mag.) and I usually find it in the triangle formed by β Ceti, Grus and the 1st mag. Fomalhaut, with Sculptor at the same altitude as Fomalhaut and 30° east of it. Sculptor contains two major astronomical objects, one of which is visible in binoculars. **NGC55** is a famous edge-on galaxy and one of the nearest outside our local group. Unfortunately, at 8th magnitude, it is just beyond visibility for binoculars. But then... there's **NGC253**.

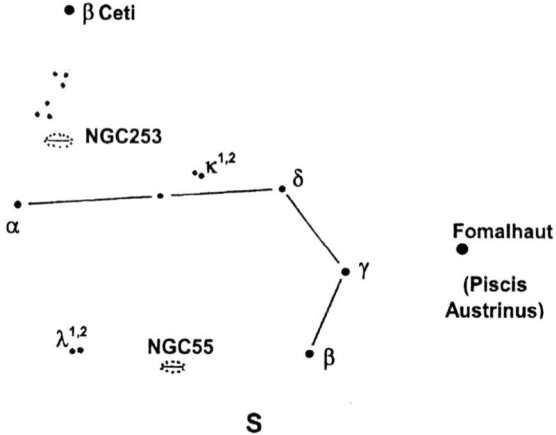

NGC253 can be found by moving approximately 7° south of β **Ceti**, the 2nd mag. and brightest star in Cetus towards α Scupltoris. Conveniently there are two triangles of 6th magnitude stars (visible in binoculars) that point the way to the galaxy (see the map). Go from β Ceti. towards α Scl., find the triangles and the galaxy is **there.**

NGC253 is a magnificent edge-on spiral galaxy about 9 million light years away containing hundreds of billions of stars, and at 7th mag. can be seen as a cigar shaped smudge of light, about as long as the Moon's diameter. It is the next most visible galaxy (to binoculars) after the great Andromeda Galaxy M31. Make sure your eyes are well light adapted, as in all but the darkest country skies it is the faintest of smudges.

An easier target is found 1/3rd the way from δ to α **Scl.**, 1° off the line. It's the wide optical double κ¹ and κ² **Scl**. About 0.5° apart, of mags. 5.5 and 7, their colours vary greatly, κ¹ being a yellow-white 225 light years away (and a true binary star itself, too close for binoculars), and κ² being an orange about 1,000 light years away.

Find α **Scl.** and move south about 10°. You'll find another wide double (about ¼° apart), λ¹ and λ² **Scl**. They are of similar magnitude (6th) and distances (390 and 326 light years) but completely different classes. λ¹ is a blue-white giant while λ² is an orange star. Can you see the colour differences?

Scutum – the Shield (June to October)
(High northerly aspect)

A small kite shaped constellation of faint stars, it can be located on the monthly star map north of Sagittarius about mid-way between Scorpius' tail and the bright star Altair in Aquila. It is the 5th smallest constellation in the sky. However, being embedded in the Milky Way it contains an area of particularly rich star fields, well worth scanning with binoculars.

Scutum's pride is **M11**, a cluster of stars called **'The Wild Duck'** cluster, comprising over 200 stars, 6,500 light years away. It is found about 2° south-east of β Scuti. It is in fact an open cluster, but because of its compactness (about ½ a Moon diameter) it appears in binoculars like a misty globular cluster. It gets its name from the appearance like a large fan shaped flock of ducks in flight.

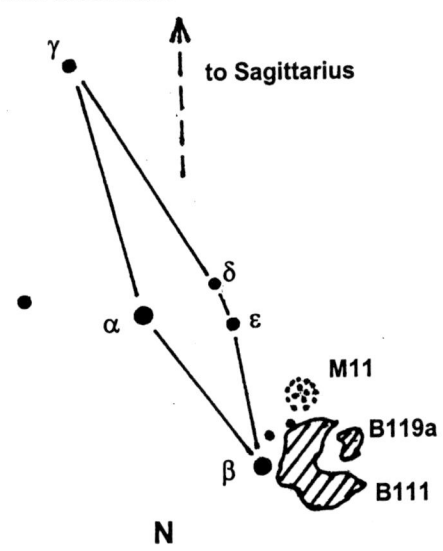

Find ε Scuti 2° east of α Scuti. ε is an attractive wide optical double, comprising a mag. 5 yellow star 750 light years away, and a mag. 7 orange star 650 light years away.

For something different, aim your binoculars north of **M11** so that **M11** and **β Scuti** are both in your field of view. What do you see? If you see a lot of nothing, that's the Dark nebula called **Barnard 111**. It's about 2° long, with an L shape at the north end, pointing east. Like the Coal Sack in Crux, it obstructs light from the stars behind it.

Tucked just south of the eastern leg of **Barnard 111** is another smaller dark nebula, **Barnard 119a**, a crescent shaped patch about 1° long. Binoculars are ideal for viewing dark nebulae like these, especially in dark skies where the richness of the Milky Way is visible. In fact, these dark nebulae are unlikely to be visible unless you have a good dark sky, free from light pollution.

While you are touring Scutum, you are likely to bump into a number of other clusters as γ Scuti borders on the binocular rich area of Sagittarius.

Serpens – the Serpent

(Northerly aspect) (May to August)

Imagine a snake cut in half by a train. To the left of the track, the head. To the right, the tail. That's the constellation Serpens, two parts separated by Ophiuchus in the middle, around whom the snake is wound. The stars of Serpens (west – Caput, the head, east – Cauda, the tail) are faint 3rd and 4th mag. but can be found relative to Scorpius, Sagittarius and Ophiuchus.

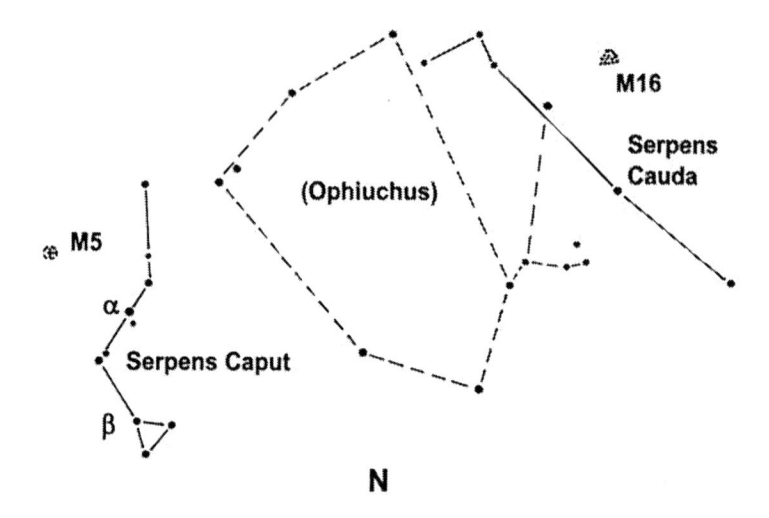

Each of the Head and Tail have an object of binocular interest.

To the west of Ophiuchus, find α **Ser**, an orange mag. 2.7 star. Move 8° SW and find a 6th mag. fuzzy ball, just on (or beyond) naked-eye visibility. This is **M5**, a globular cluster 26,000 light years away. This is a magnificent object, arguably the 5th brightest globular in the sky. You can add this to your list of favourite globulars.

Then jump east across Ophiuchus to the serpent's tail. We are looking for a fainter (mag. 8, close to the limit of binoculars) object, **M16**, an open cluster contained within the famous Eagle Nebula 16,000 light years away. It is best found by locating Sagittarius, then **M17** in Sagittarius. (See earlier section). Go 3° north of **M17** and see a faint hazy cluster of stars. (You'll need a good dark sky for this.) Binoculars will give you the star cluster, but that's only a mere hint of the glorious object the cluster is contained in – the Eagle Nebula. This nebula must compete with the Orion Nebula in telescopes for visual beauty. All you will see of the nebula in your binoculars is a suggestion of mistiness around **M16**.

Taurus – the Bull (November to March)
(Northerly aspect)

This is one of the great constellations from Greek mythology. It can be the great white bull that Zeus disguised himself as to carry off Europa, or the bull put up in the sky by Zeus for Orion to fight after he died. The face of the bull is easily seen as the big 'V' in the sky, with the point of the 'V' as its nose, the open end of the 'V' as its eyes. The tips of the horns are a long way off, the stars β and ζ. Taurus was designed for binoculars – enjoy!

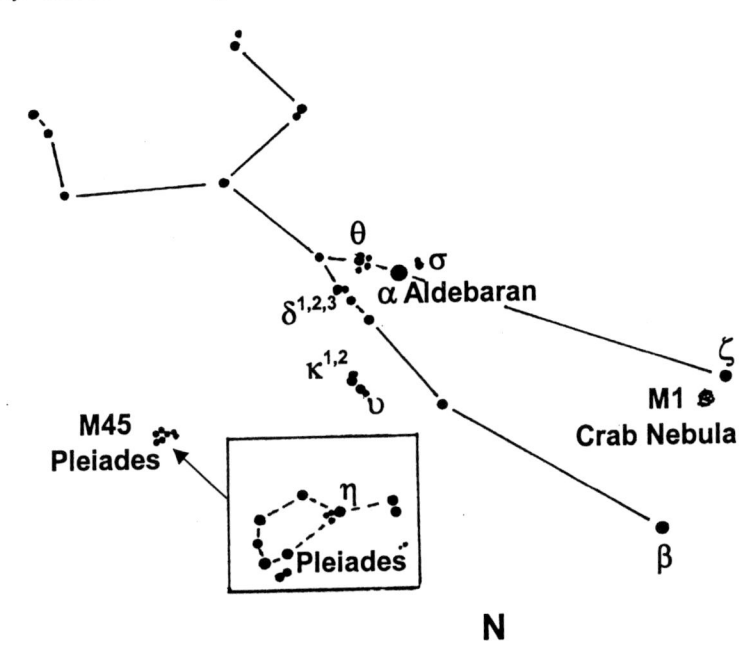

Let's start with the big 'V' itself. This easily identified arrangement of stars has the name **Hyades**. Believe it or not, with the exception of the red giant star, all these are part of an open cluster only 150 light years away, and are the second nearest open cluster to us. This makes them a very important tool to astronomers for measuring the scale of the Universe. The blazing red giant star, the bull's eye, is Aldebaran, the 14th brightest star in the sky. It is not really a part of the cluster, being only 65 light years away.

On the southern (top) limb of the V there are two naked-eye (and therefore binocular) double stars, 5' apart. The first, θ **(theta) Tauri** is nearest the nose and contains an orange (mag 3.9) and a white (mag. 3.4) star, both giants.

Then just 1° south of (above) Aldebaran is another naked-eye/binocular double star, σ **Tauri**. Both are white stars, mags. 4.7 and 5.1, 7' apart.

On the opposite side of the V, 5° below (north) of Aldebaran, is a faint 4th mag. star, κ **(kappa) Tauri** that forms a binocular double of two white stars about 0.1° apart.

Only 0.5° north of κ is another optical double, comprising ν and **72 Tau**, a wide 17' apart with a mag. 5.5 blue and a mag. 4 yellow-white. Then, also on the V but directly opposite θ you'll find a trio arranged like a flat isosceles triangle about ¾° long. It's δ^1 (an orange), δ^2 (white-yellow) and δ^3 (white) Tau. Scan in and around the V carefully and you will observe a number of other star groupings which are not formally listed as double but are a pleasure to see.

Then there is the show stopper, a major **Wow!**, the **Pleiades**, also known as **M45**. (But how Messier could ever have thought of this as a comet candidate, I'll never understand.)

This star cluster is so prominent and eye catching below and to the west of Hyades that it pops up in the legends and folklore of nearly every culture in the world, back to remote antiquity, including the Australian Aboriginals.

The naked eye can see from 6 to 9 stars in the group, depending on your eye and the darkness of the sky. The ancients used it as an eye chart. If you could see 7 stars, you had 20-20 vision. The Greeks named the seven main stars after the nymph daughters of Atlas and Pleione. The two stars at the tip of the handle are the proud parents themselves. To be totally different, the Japanese describe them as *Subaru*, meaning 'a string of jewels' and they appear in stylised form on the badge of that car. However, Pleiades is actually a cluster of about 200 stars and are all about 380 light years away. They are mostly 'new born' giant blue-white stars, less than 50 million years old. That is, they are very young, very large, and very, very hot. Like the Hyades, the Pleiades are best viewed through binoculars.

Some people claim they can see the **Crab Nebula (M1)** in binoculars. Theoretically this is possible, though it would be maddeningly faint. Find the bull's horn tip ζ **Tauri** about 15° from Aldebaran. M1 is about 1° NW of ζ. Expect to see, if anything, the faintest of nebulosity (averted vision would be useful here), the remnant of a star that exploded in 1054 AD, and is 6,500 light years away.

Why do stars twinkle?

It is due to the effect on the starlight by the Earth's atmosphere. The image from a star, being so far away, is effectively a single path of light photons that should hit only one spot on your eye. However, the varying layers of moving and moist air it passes through causes individual photons to strike different parts of your eye, making it appear to jump about, or twinkle. Seen from space, stars would not twinkle.

Telescopium – the Telescope (June to October)
(High overhead, southerly aspect)

One of the uninspiring constellations devised by the French astronomer Nicolas de Lacaille which only demonstrates his overactive imagination. However, for binoculars it does contain a pretty optical double. It is easiest found tucked midway between Corona Australis and Ara.

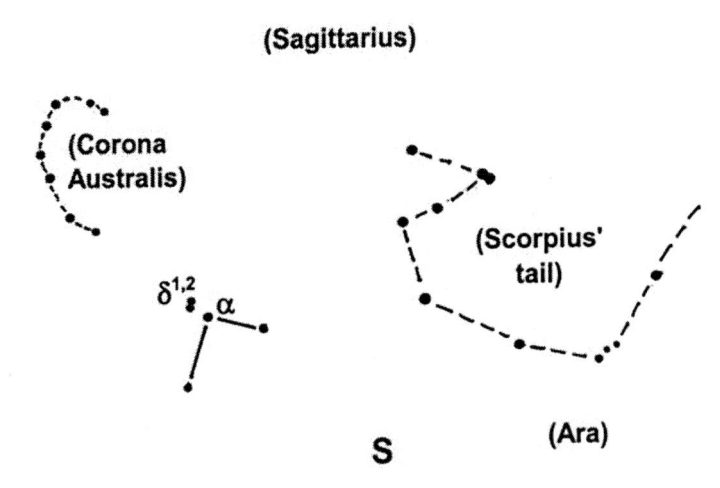

Just 1° east of the apex of the right angled shape (made from stars of magnitudes 3.5 to 4.5), there is the wide (9' apart) optical double star, δ **Telescopii**. Its components δ^1 and δ^2 are both blue-white stars of magnitudes 4.9 and 5.1, very similar in brightness. However, they are 800 and 1,120 light years away respectively.

Why don't planets twinkle?
Light from a planet suffers the same "jumping about" effect on your eye from the atmosphere as the light from a star. But a planet is immensely closer and the light from its surface disk comprises a number of parallel paths of photons, not just one. Each path jumps about on your eye but the combined simultaneous jumps of all the paths cancels the effect out and the image appears steady. That is, it doesn't appear to twinkle.

Triangulum – the Triangle (November to January)
(Very low northerly aspect)

Triangulum is a small constellation (about as long as the Southern Cross but very narrow) marked by three stars (two at mag. 3, one at mag. 4) in a very acute angled triangle shape. It is found about midway between Pleiades and the Andromeda Galaxy M31, or about 25° east of the bottom right star of the Square of Pegasus (also the head of Andromeda). Also, it can be found 10° directly below the three main stars of Aries. Three ways to find a triangle seems appropriate.

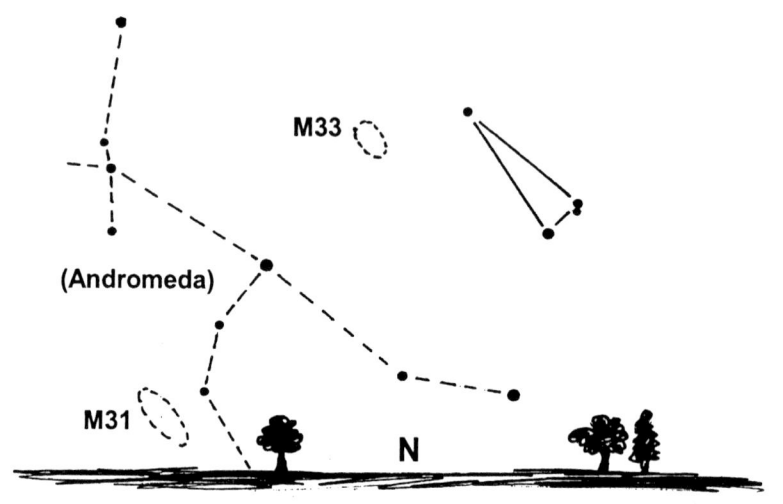

A prime binocular object in Triangulum is **M33**, the **'Pinwheel Galaxy'**, found about 4° west of α Trianguli. This is a face-on spiral galaxy 'only' 2.7 million light years away and a member of our local galactic group. If you have a dark site, this is better seen in binoculars rather than a telescope because of its large size (larger than the Moon) and low brightness. You will note that **M33** does not have as concentrated a centre as most spiral galaxies. Look for a large area of dim glowing light, a bit larger than the size of the full moon. It will not have any area of bright or concentrated light and averted vision will be very helpful.

This is a classic case of if you didn't know exactly where it is supposed to be, you wouldn't notice it.

If you can see it, remember ... around 400 billion stars as they were 2.7 million years ago. ***Wow!***

You will notice that the bottom right hand star of the triangle is a wide double (22' apart), comprising a 4th magnitude white star(γ Tri., 150 light years away) with a 5th magnitude yellow optical companion (δ Tri. 33 light years away).

Triangulum Australe – The Southern Triangle
(Low to high southerly aspect)   (All year, but highest April to September)

An equilateral triangle about as long as the Southern Cross, found on the opposite side of the Pointers from the Cross. It represents a carpenter's tool, the Triangle. Not very mythical, but a lot of the southern constellations are like that. Its stars are reasonably bright from mag. 1.9 to 2.9.

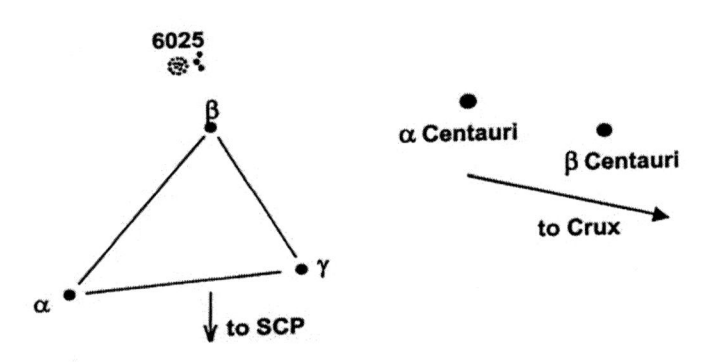

The binocular object is **NGC6025** found about 3° north of β **Tr.A.** Alternatively, extend the line from β to α Centauri about 2.5 times and it will be just under that point. It sits cosily beside a trio of stars shaped like a boomerang. It's a 5th magnitude open cluster 2,500 light years away containing about 60 or so 7th magnitude stars, easily visible in binoculars and just naked eye visible from a dark sky. To me it suggests an oval or tear shape.

It's a good idea to check out this constellation and its cluster at the same time as the clusters in **Norma**. They abut each other and **NGC6025** is very close to **NGC6087** (only 3° away).

How old is the Universe?

This is generally asked in the context "How long is it since the Big Bang?"

Once believed to be between 10 and 20 billion years, precise measurements of fluctuations in the microwave background radiation give a confident figure of 13.7 billion years, plus or minus 200 million years. If you don't accept the Big Bang model, you can ignore this figure.

Tucana – the Toucan (All year, but highest August to January)
(Low to high southerly aspect)

A faint constellation of 3rd to 4th magnitude stars in the less well known part of the sky close to the South Celestial Pole (SCP). Representing the colourful large beaked bird from South America, this constellation was created by the late 16th century Dutch navigators Keyser and de Houtman. The easiest way to locate it is to find the SCP (see the earlier section on finding the SCP) and find the spot directly opposite and the same distance as the Southern Cross. Tucana contains a number of very interesting binocular objects.

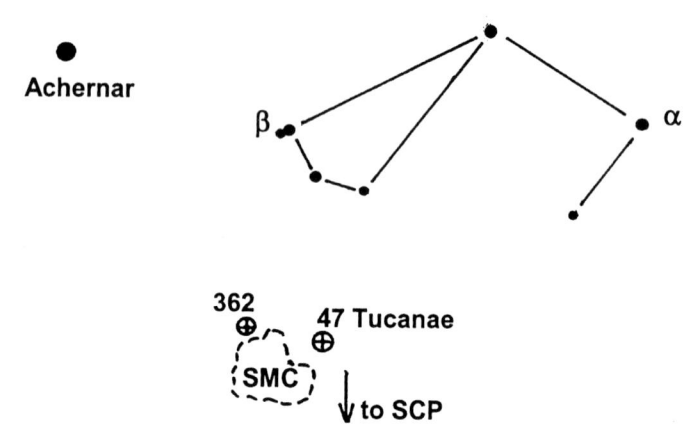

The eastern-most star, β **Tuc**, found 9° south-west of Achernar, is a multiple star but is seen only as a binary in binoculars. It appears as two very close (0.5') near identical 4th mag. white stars (very steady and larger binoculars required here). β² is in fact a close binary needing a telescope to split. There is another unrelated optical companion, a 5th magnitude white star 9' from β.

Then move 10° towards the SCP to find a gathering of three deep sky objects. The largest is the **Small Magellanic Cloud (SMC)**, seen by the naked eye (at a dark site) as an elongated wisp of cloud about 7 Moon diameters long. Study its area with binoculars to find patches of nebulae and star clusters. SMC is a nearby satellite galaxy, classed as an 'irregular galaxy', only 190,000 light years away.

Immediately beside the SMC to the 'west' is **47 Tucanae**, an absolutely beautiful globular cluster, second only to ω Centauri. You should be able to see it without binoculars as a fuzzy 4th magnitude star and so is easily found in binoculars. Compare it to ω Centauri for brightness and central concentration. While gazing at 47 Tucanae, remember it is 16,000 light years away while the adjacent SMC is over ten times further away.

Tucana has another globular, the 6th magnitude **NGC362**, located at the top (north) end of the SMC. At 29,000 light years distant, it is obviously not as bright or spectacular as 47 Tucanae, but still fun to hunt down.

Vela – the Sails (January to June)

(High southerly aspect)

Vela is the other third of the old constellation Argo Navis, the rest made up by Carina and Puppis. It is large, about 40° by 15° and lies in the thick of the Milky Way. Vela is most easily found as the large area above the top right side of the False Cross. In fact, the top and right hand stars of the False Cross are κ (kappa) and δ (delta) in Vela.

Vela abounds with binocular clusters plus a binocular double.

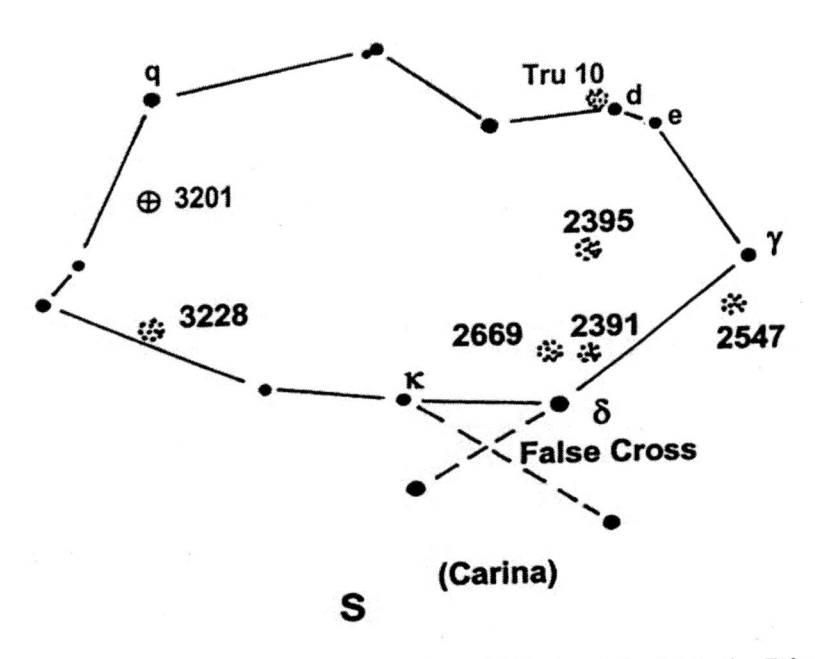

Let's start with the star γ **Velorum**, about 10° to the NW of δ in the False Cross. γ is a multiple star but binoculars will show a close (42") optical double comprising mag. 1.8 and mag. 4.3 blue-white stars. Steady binoculars are needed here.

Move 2° south of γ and find a 5th mag. open cluster **NGC2547**, about 1,400 light years away and containing 80 or so 7th magnitude stars.

Jump down to δ **Velorum** in the False Cross. 2° north of δ you will find **IC2391**, a large open cluster of about 50 stars, approx. 500 light years away clustered around a brighter 3rd mag. blue-white star.

Then about 1° to the east you will find a 6th mag. open cluster called **NGC2669**.

If you sweep 5° due north of **2669**, you will find **IC2395**, 3,100 light years away, a 5th mag. open cluster of about 40 stars.

Then move up (north) from **2395** and locate the two 4th mag. stars of **d Vel.** and **e Vel**. Shown on the map immediately to the east of **d Vel.** you will find a field of stars. Embedded in that field is a loose knot of about 40 brighter stars called **Trumpler 10**. It covers an area approximately as large as the Moon. This cluster is 1,100 light years away with stars as young as 36 million years.

Moving to the eastern side of Vela, approximately one 'False Cross length' beyond the top of the False Cross, find **NGC3228**, an open cluster 1,600 light years away, containing a small group of about 15 stars. Then, from **3228,** move north by about **5°** halfway to the star **q Vel.**, and you will locate the globular cluster **NGC3201**. At magnitude 6.8 and 16,000 light years away, it is visible (from a dark site) as a fuzzy blob. **Note:** Like snow flakes, no two open clusters look exactly alike. When you are finding these clusters in Vela (and in other constellations), see if you can identify the differences.

Measuring a star's distance by parallax.

If you hold up a finger at arm's length with one eye shut, then with the other eye shut, you will see your finger appear to move against the background. That is an example of parallax. Astronomers used the same method to measure distances to the stars up to about 300 light years away. But instead of the baseline being the distance between two eyes, they used the distance between the Earth's position at opposite sides of its orbit around the Sun – about 300 million km. The angle subtended by half this distance is called the star's 'parallax'. Even for the nearest stars, this is a very small angle, less than one second of arc (1/3600 of a degree). A star with a parallax of 1" would have a distance of 1 parsec, which is 3.26 light years.

The first star to have its parallax measured was 61 Cygni by Bessel in 1838. The satellite Hipparcus, orbiting a long way out from Earth and not having refraction problems with Earth's atmosphere, has measured the parallaxes of hundreds of thousands of stars to accuracies of better than 0.001" arc. We now know the distances to these stars very accurately, leading to great advances in astronomical science.

Vulpecula – the Fox (July to October)
(Northerly aspect)

The stars of Vulpecula are very few and faint. The best way to describe its location is 'under the arrow' of Sagitta, between Sagitta and Lyra. But ordinariness can be deceiving as Vulpecula contains two fascinating binocular objects. One is famous but difficult to find. The other is very easy and, as a fitting end to this book of **Wow!s**, a great party trick.

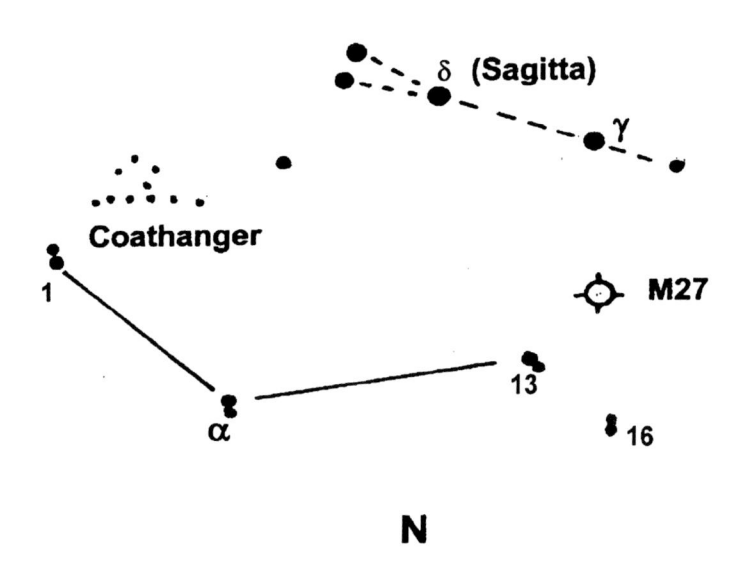

First an easy optical double. When you locate the shallow V shape of Vulpecula (from the bottom of the map), the star at the bottom of the V is the brightest at mag. 4.4. This is α **Vul**. It consists of a red giant star 399 light years away with a wide optical companion, an orange giant at mag. 5.8 which is 480 light years away.

Then look at the stars **1 Vul.** and **13 Vul.** at each end of the shallow V. They too are wide optical doubles comprising 5th magnitude white stars. In the vicinity about the shallow V, there are many other wide optical doubles. An example is **16 Vul.** with a 9' separation of a 5th mag. yellow and 6th mag. white star.

The challenging object is **M27**, the famous **Dumbbell Nebula**. To find it, first identify γ **Sagittae**, the brightest star in Sagitta, the tip of the arrow. Then approx. 3° north of γ is **M27**. Another way is to find δ, the centre star of the arrow, and swing it down around γ **Sagittae** like a pendulum for 120° ($1/3^{rd}$ of a circle). This will also bring you to **M27**.

(Notice that all these tips involve stars in Sagitta, not Vulpecula. This is because the stars of Vulpecula are so faint and difficult to identify. It's easier to use Sagitta and you can ignore Vulpecula if you so wish. However, if you don't so wish, go up 2° SE from 13 Vul. which you found earlier and this should land on M27.)

M27 is an 8th magnitude object 1,000 light years away, just visible in 50mm binoculars but you will need a dark site. It is a planetary nebula and famous for being the most conspicuous of its type in the sky. In a telescope, it actually looks like a dumbbell, not like the classic circular smoke ring we expect from planetary nebulae. In your binoculars, you will see a misty smudge, with a hint of a dumbbell (or hour glass) shape. It's another case of if you have seen the photos, you'll know what you are seeing.

The other object of interest, the 'party trick', is the so-called **'Coat Hanger'**. It has a name (Collinder 399 or Brocchi's Cluster) but it will always be known as the Coat Hanger. This will tickle your funny bone – a real cosmic joke. First you need to find Sagitta again. Find the two tail feathers of the arrow and move about one arrow's length north-west at an angle of 60° off the arrow's line. You will find, very clearly in your binocular view, a set of faint stars that looks very much like a simple coat hanger. Six stars in a dead straight line, with a large hook protruding upwards from the centre of the six. If you can't find it first off, don't give up. It's there, about 60° off the arrow's line about one arrow length behind it. The Coat Hanger is approximately 1.5° long and 0.5° tall, so you can't miss it.

This object has no astronomical significance. It is an asterism, a fluke arrangement of unrelated stars at vastly varying distances. Like cloud gazing, it's fun to just scan the heavens and see odd shapes. But this one makes you wonder – who hangs his coat on it?

On Reflection...

So, you've come to the end of the tour. You probably will have found some objects a lot more easily than others. Some you may not have found at all, for a variety of reasons. Definitely you will have found some that caused you to think, if not say aloud, **Wow!**

Hopefully I will have enthused you enough to go back to these from time to time to enjoy them over and over again, always keeping that sense of awe about their place in the size and complexity of the Universe, not to mention your trip in the Time Machine.

Also, I hope you will continue to try to track down those objects you may have had difficulty finding, perhaps waiting for better conditions at a darker site.

And, finally, I trust that your experiences, your sense of Wow!, will encourage you to share the fun with your family and friends so that more people may have the opportunity to explore and understand this wonderful Universe in which we live.

Clear skies!

Robert Bee

About the Author

Robert Bee lives at Mount Annan on the south-west outskirts of Sydney.

Robert's passion for astronomy began in his teens and has deepened over the ensuing years. With degrees in Electrical Engineering and Science, he enjoys both observing the starry sky and understanding the physical laws behind what he sees.

Robert is a member of the Macarthur Astronomical Society (MAS) and has edited and contributed to the Society's monthly journal "Prime Focus" since it commenced in 1996.

He shares his passion for astronomy with the people of the Macarthur Region through a fortnightly column called 'Heavens Above!' in a local newspaper. This column commenced in July 1998 and is aimed at those with no background in science or astronomy, just a sense of curiosity and a willingness to step outside the back door and have a look.

Robert also enjoys writing fiction, with a preference for science fiction and fantasy, and has had a number of short stories published in periodical magazines and successes in short story literary competitions. He currently has a children's science fiction novel, with an astronomy theme of course, in progress.

Robert enjoys talking to the public about astronomy and guiding them around the sky, both at public nights run by MAS and also at clubs, societies and schools.